The Search for India's Rarest Birds

Edited by
Shashank Dalvi and
Anita Mani

🌀 juggernaut

JUGGERNAUT BOOKS
C-I-128, First Floor, Sangam Vihar, Near Holi Chowk,
New Delhi 110080, India

First published by Juggernaut Books 2024

Anthology copyright © Shashank Dalvi and Anita Mani 2024
Copyright for the individual essays vests with the respective authors
Pages 265–66 are an extension of the copyright page

10 9 8 7 6 5 4 3 2

P-ISBN: 978-93-5345-320-6
E-ISBN: 978-93-5345-725-9

The views and opinions expressed in this book are the author's own. The facts contained herein were reported to be true as on the date of publication by the author to the publishers of the book, and the publishers are not in any way liable for their accuracy or veracity.

All rights reserved. No part of this publication may be reproduced, transmitted, or stored in a retrieval system in any form or by any means without the written permission of the publisher.

Typeset in Adobe Caslon Pro by R. Ajith Kumar, Noida

Printed at Thomson Press India Ltd

For Ramki S., who was Shashank's crime partner while they chased near mythical ('legendary') birds in the mountains of Northeast India and beyond!

and

For Maitreya, who first raised Anita's eyes skyward to the beauty of birds.

Contents

Introduction		1
1.	A Charismatic Duck Painted in Carnation and Chocolate	10
2.	The Rediscovery of the Jerdon's or Double-banded Courser	28
3.	A Tale of an Absconding Owl	58
4.	The Chilappan Challenge	75
5.	Following in the Footsteps of the Elusive Masked Finfoot	98
6.	Owl of the Emerald Island	114
7.	In Search of the Last Megapodes	130
8.	Nong-in – The Bird that Tracks the Rain	151
9.	In Search of Vicky on Phawngpui	165
10.	Lusting for a *Locustella*	181

11. Twitching Tales 196

12. Finding the Next Rare Bird for India 221

Notes 229

List of Non-human Animals Mentioned in the Book 252

Acknowledgements 261

A Note on the Editors 263

About Indian Pitta 264

Copyright Acknowledgements 265

Introduction

The biggest high for a birder is coming upon a species hitherto unknown to science. When faced with a species that defies every description in a field guide, with a call unrecognized by mobile apps like Merlin, or online sound repositories such as Xeno-canto, what do birders do?

There is rising excitement (which they must fight lest they are being led on by a mischievously cloaked juvenile bird), as images and calls are shared with ornithological savants (often live from the field, over WhatsApp and other social media networks), then exultation, as the return messages indicate that they are on the precipice of a rare find, a new bird.

Alas, in India, where ornithologists, naturalists, collectors and hobbyist birdwatchers have criss-crossed the subcontinent's myriad habitats and landscapes for over two centuries, this moment does not arrive often. The British Raj was a period of enormous discoveries – there were new bird species to describe and information to collect on species seen for the first time. This resulted in the first-recorded baseline data for species distributions.

To date, this information is still being used in our field guides. Much of the credit for this goes to naturalists such as A.O. Hume, T.C. Jerdon, Edward Blyth, Brian Hodgson, Samuel Tickell and John Gould (who, surprisingly, never set foot in India) who were active between the mid-1800s and early 1900s.

In fact, the early ornithologists did such a thorough job that just five new species have been described since India's Independence in 1947 – the Mishmi Wren-Babbler (by Dillon Ripley in 1948), the Nicobar Scops Owl (Pamela C. Rasmussen, 1998), the Bugun Liocichla (Ramana Athreya, 2006), the Himalayan Forest Thrush (Per Alström et al , 2016) and the Ashambu Sholakili (V.V. Robin et al, 2017). But there is still plenty of excitement to be found in the pursuit of species that have been lost for decades. As James Eaton writes on his quest to find long-lost birds, '… that feeling of laying your binoculars on the prize is an addiction greater than what any narcotic could give you.'

In the last 40 years, there has been a surge in scientific ornithological studies as well as hobby birdwatching, both of which have fed off the availability of the tools of the trade. Before 1998, the most reliable source of information was Dr Sálim Ali's and Sir Dillon Ripley's 10-volume *Handbook of the Birds of India and Pakistan*, published between 1968 and 1974. One couldn't carry these tomes to the field, so identification became an exhausting process. Then came modern-day field guides by Grimmett, Inskipp and Inskipp (1998), Krys Kazmierczak (2000) and Pamela C. Rasmussen and John C. Anderton (2005, 2012), which made bird identification easier. All three were easy-to-use and included updated information. In addition to these, the creation of e-groups, such as bngbirds and birdsofbombay, allowed for the exchange of information over the Internet. Webpages such as www.kolkatabirds.com and www.orientalbirdimages.org served as extremely useful databases. Simultaneously, two developments in India in the early 2000s – the IT boom and the arrival of digital cameras – triggered a massive surge of interest in birding and bird photography. Over the past decade, birding has become even more digital in India, with the appearance of recording equipment, which had been an integral part of birding in Western countries since much earlier. Today, websites such as www.xeno-canto.org are extremely popular among birders, both for locating birds in the wild and for identifying them. Over the last 15 years, platforms such as Facebook have catalysed the creation of many birding and wildlife groups; examples include Indian Birds, with over half a million members, and Sanctuary Asia, with over 2,50,000 members. Today's birders also have recourse to powerful global platforms, such as eBird, to track and maintain records of species sightings.

Given this increasingly supportive environment, species that had been missing for decades have been rediscovered in the last 30–40 years. This book chronicles the stories of those finds. The process of rediscovery is organically different from new species discoveries. The latter is often accidental, and may occur while exploring the path less trodden – as it happened in the case of the 1995 sighting of the Bugun Liocichla by Dr Ramana Athreya – or when the birdwatcher chances upon a cryptic species with a brand new song, as in the case of the Himalayan Forest Thrush. Per Alström and Shashank Dalvi knew for certain that the thrush they were looking at was singing a very different tune (literally) as compared to the more familiar Plain-backed Thrush. Years of genetic and museum studies later, in 2016, the different-sounding song led to the description of the Himalayan Forest Thrush as a new species to science.

Rediscoveries, in contrast, are often premeditated, though the trigger may well be a chancy find, as Pamela C. Rasmussen's tale of the rediscovery of the long-lost Forest Owlet illustrates. An odd listing in the register of a century-old collection drew her down a path of enquiry, which led to the realization that birdwatchers had been looking for the bird in the wrong place for decades. But it took hundreds of hours of meticulous work with museum specimens coupled with fieldwork, for the pieces to fall into place. The study of museum specimens is crucial for delineation, especially of cryptic species that can be decoded only in the hand. The specimens can also provide valuable clues for the search – enlarged gonads or a brood patch, if present, indicate that the specimen was collected during the breeding season. This may point the seeker towards a specific period of the year – the breeding season in this case – when it might be easier to spot the bird, as they tend to be more conspicuous or vocal.

Though the tools of rediscovery – museum specimens, fieldwork and, increasingly today, genetic analysis – are the same as that for the discovery of a new species, the scenario has changed dramatically from the nineteenth and twentieth centuries, when so many new species were described. Field expeditions of the past required travelling to remote locations and often lasted months. Today, roads have reached the remotest corners of the country, and while the road is a metaphor for habitat destruction, it has made rediscovery a lot easier. Roads have also made weekend twitches – when you parachute into a remote location to tick-off a species for your life list – possible, as Atul Jain's tale of his twitching adventures illustrates.

Roads, a metaphor for development-led destruction of a habitat, are also the reason why some searches remain fruitless. Aasheesh Pittie's deep dive into the written history of the Pink-headed Duck in India (for that is all we have of what he calls 'the seemingly disdainful Pink-headed Duck') was nothing short of an exercise in literary forensics to discover why this bird has eluded so many searchers. It becomes clear that the species was rare in its distribution to begin with, and was possibly common only locally within its wider range, which in the nineteenth and early twentieth centuries extended across northern and eastern India (with some records from the south as well) in the winter, with a narrower breeding range restricted to parts of Bengal and Bihar. The press of human activity was calamitous for such a species. Sadly, as the cases of the White-bellied Heron and Great Indian Bustard demonstrate, history seems ready to repeat itself.

In today's world, even if roads exist, it is the birder who makes the effort that reaps the rewards. As Andrew Spencer and Puja Sharma's story of the finding of the Mount Victoria Babax on Phawngpui – Mizoram's highest peak – shows, all clues indicated that the bird was likely found there, but it took a determined duo to commit time and make a dedicated search to find the bird. Sayam Chowdhary's quest for the Masked Finfoot in the myriad waterways of the Bangladesh Sundarbans or Praveen J.'s search for the Banasura Laughingthrush similarly show that finding or rediscovering a bird requires strategy (where to look) and stamina (read: relentless fieldwork). In many cases, this is a race against time as rare birds are often clinging precariously to the few remaining pockets of their habitat – the Bangladesh Sundarbans is one of the last strongholds left for the Finfoot, while the Banasura Laughingthrush is in an equally precarious position, stranded as it is at the mercy of a very restricted elevational range and climate change.

Make Mine Rare

What makes a bird rare? The immediate meaning points to birds found in low numbers and within a specific type of habitat. The Jerdon's Courser would certainly qualify as one. As Bharat Bhushan's exciting search for this bird shows, the ground dweller was found in a narrow band of 'plains-forests below the hilly regions', and more specifically, as Bhushan states in his essay, within 'thin and open scrub near rocky hills'. Even within such a precise range, sightings were sporadic.

But paucity of records has sometimes more to do with accessibility or the explored status of a habitat rather than species numbers as demonstrated by the Nicobar Megapode and Nicobar Scops Owl. The logistics of travel to the Nicobars is complicated, but once you get there, a concerted search should reveal the Megapode, as Radhika Raj's essay on this endemic species demonstrates, while the owl is surprisingly hard to miss, as Shashank writes. Also, take, for example, the Mrs Hume's Pheasant, which was a mythical bird for birders, until recently. But increased explorations of the Naga Hills – close to the Manipur–Myanmar border and further north into Nagaland have shown that given the right habitat (oak conifer mosaic), time of day (a nighttime search is fruitful) and specific skill set (to see the bird you have to enlist local hunters turned bird guides whose tracking skills are unparalleled), a birder is highly likely to see the bird. The same is true of the once mythical birds of the North-East; in early 2000s, most such birds – Yellow-throated, Brown-capped and Moustached Laughingthrushes, Gould's and Rusty-bellied Shortwings, Mishmi Wren-Babbler, Hodgson's Frogmouth, Sikkim and Cachar Wedge-billed Babblers to name a few – could only be seen in field guides, while some required an expensive birding trip to Bhutan. Today, the access provided by roads, tourism infrastructure and excellent bird guides means that many once rare birds can be seen on a well-planned trip.

Access has also levelled the playing field. With the great surge of interest in birdwatching, many more Indians are, literally, taking to the field. This means that many of these rediscoveries are being made by Indian ornithologists, naturalists and birdwatchers, thus passing the ownership of knowledge into Indian hands. In the past, species discoveries were part of a larger pattern of intellectual colonization. It also didn't help that in the pre-Internet era, scientists and biologists from well-funded institutions, typically from the West, had access to the subcontinent's museum specimens and printed information which were not available on the subcontinent. In a telling example, foreign biologists described a group of lizards from the Andaman & Nicobar Islands as new species to science, based on collection done by Indian biologists; the former claimed the credit and the glory of discovery without setting foot on the islands while, sadly, the latter were not even credited as co-authors and their contribution was simply glossed over in the 'Acknowledgements' segment of the manuscript. It is another matter that the same pattern is being repeated by Indians scientists, who 'scoop' data and discovery work done by peers – there are many such examples in the world of Indian herpetology, which is the latest hotspot for new species discovery. Notwithstanding the irony, we are, as Indians reclaiming our natural heritage.

Fortune Favours the Brave . . .

. . . And the intrepid explorer who is willing to go boldly where no birdwatcher has gone before. Chasing the next set of rare birds for India – and there are some exciting possibilities, as Frank Rheindt lays out in the concluding essay of this book – will require birders to move out of their comfort zones and chart the unchartered. To make this happen, they will have to disprove what James Eaton says about them – 'Birders generally are a bunch that enjoy treading the well-trodden path, rarely straying to explore new areas.'

It is also time to go beyond discovery to study the ecology of a species post rediscovery. The raft of scientific work that followed Rasmussen's rediscovery of the Forest Owlet has few parallels. Other kinds of follow-throughs may also be absent. As Bhushan comments on the 're-disappearance' of Jerdon's Courser, '... the species has not been reported or sighted elsewhere in its known range, because of the lack of appropriately methodical surveys.'

Some put down great finds to plain, dumb luck. We think differently. For the birder or the biologist willing to make the effort and stay the course, the rewards will be the greatest – for surely, the harder you work, the luckier you get!

1

A Charismatic Duck Painted in Carnation and Chocolate

AASHEESH PITTIE

The Pink-headed Duck is a species that has certainly gone extinct in the subcontinent, and we'll let the author tell you why. With no living person having seen the bird, Aasheesh Pittie performs a splendid piece of literary forensics to create a compelling picture of this stunning duck. Going beyond the piece, it is perhaps time for India to officially declare the species extinct. Such a declaration will force us to publicly recognize a reason for its downfall – habitat loss. With the fate of birds, such as the Jerdon's Courser, Great Indian Bustard, Lesser Florican, White-bellied Heron and Finn's Weaver also hanging in the balance, there is no better time to own up to this, and act to fix it. If the Pink-headed Duck could motivate us to protect key habitats and come up with species-specific conservation plans and their implementation, its loss would have not been in vain.

Bhawani Das studied the bird intently. Most of the time it stood stock still in the large aviary, occasionally preening, listlessly pecking at the ground, or rheumatically waddling a few yards, but never far from him. It was a bright winter morning sometime between 1777 and 1782,[1] and Das could not complain about his subject. It had clean lines and a simple colour pattern. He noted the swan-like neck, the sloping forehead and the reddish eye fixed upon him. He had been commissioned to make its portrait.[2]

Das hailed from Patna (Bihar) and had been trained in the style of the Mughal School of miniature painting. But the group in Patna had developed their own identity by merging the styles of Mughal and European schools, forming the Patna School or Patna Qalam, often using English paper and watercolours for their work.[3] Unlike most such schools of that time, patronized by royalty and the landed gentry to glorify themselves and their achievements, artists of the Patna Qalam painted quotidian dramas they saw around them, ethnographic portraits of natives engaged in various trades and occupations, and the natural environment, specializing in native flora and fauna. They did not embellish their work with illuminated borders or illustrated backgrounds – simply painting their subjects against a plain white setting, thereby forcing the viewer to focus on the subject by eliminating every other element that would distract from it. They used natural pigments for colours and specialized in creating delicate shades in watercolour.

Work was scarce in the late eighteenth century, especially for skilled artists, and the few patrons were scattered far and wide. The competition to reach them before others did must have been fierce amongst the artists. When an opportunity to work in Calcutta (now Kolkata) presented itself, Das made haste to reach his new employers, Sir Elijah Impey, Chief Justice of the Supreme Court in Bengal and his wife, Lady Mary. The Impeys lived in a spacious house with extensive grounds. They had created a justifiably famous and impressive menagerie of native fauna in this park-like property on Burying Ground Road, which was later renamed Park Street, after the nature of their property.[4] A large aviary was part of their private zoo – outside which Das now found himself.

He picked up a brush and swirled it with a spoonful of water in a *Gulabi* (rose-coloured) pigment made from shellac,[5] laying a few tentative washes on a piece of paper he kept for testing the shades of colours. The spreading *gulabi* hue was entirely to his satisfaction, and with deft strokes he began to shape the head and neck of the bird directly onto the paper, in freehand, without having first sketched its outline in charcoal (another specialty of the Patna School). He sat comfortably, cross-legged on the ground, balancing the board with the thick art paper on his knees, the tools of his trade spread out next to him.

With each glide of his brush, the bird emerged on the paper, a close likeness of the living specimen in front of him. He shaved a larger quantity of a dark brown pigment, extracted from the bark of trees, into a shallow container and stirred in a few drops of water. With flowing strokes from a thicker brush, he filled in the body of the bird with this colour. Working swiftly with the difficult medium, he applied it in layers to darken the body and delicately highlight the feathers; then, with the finest of brushes, he elegantly created a subtle pattern on the head and neck in a uniquely Patna Qalam embellishment known as Java stippling (similar to barley grain).[6] The bare parts – legs, beak and eyes – he left for the end. So accurate was his depiction that he included a little tuft on the bird's crown, colouring it a deeper pink, revealing to ornithologists nearly two centuries later, that his model was a drake in its prime. When Bhawani Das's 47 cm high and 34 cm wide watercolour[7] was complete, its subject, a carnation-and-chocolate-coloured duck, simply titled '"Redhead" in Persian',[8] was destined to become one of the most enigmatic of birds ever found in South Asia – the seemingly disdainful Pink-headed Duck (*Rhodonessa caryophyllacea*).

Not only did the Impeys commission paintings of the flora on their grounds and the fauna in their zoo, but they also engaged talented local artists to paint scenes from the ordinary business of life that the latter saw around them, including their own little domestic dramas as subjects. In their six years of residence in Calcutta, they collected close to 300 watercolours by the likes of Bhawani Das, Shaikh Zain Uddin and Ram Das.

The Impeys 'returned to England with the pictures in 1783, and some were used by the distinguished English ornithologist John Latham (1740–1837) as the basis for his type descriptions of several Indian and Asian birds new to science.'[9] When Sir Elijah died in 1809, a part of the 'Impey paintings' was sold at auction a year later. A few were bought by the 13th Earl of Derby, a keen naturalist, for his home at Knowsley Hall, near Liverpool. It was most likely Latham who brought this unique collection to the attention of his friend, the Earl. The Liverpool Museum, National Museums & Galleries on Merseyside (UK) (now National Museums Liverpool) acquired Bhawani Das's painting of the duck from the Knowsley group in 1998.[10]

This connection, between the paintings and Latham's using them to name their subjects, is not as tenuous as it seems, for he recorded, at least for some of the new names he proposed, that he was describing them from an Impey painting,[11] or one from any of the several other art aficionados whose collections he had access to.[12] Lamentably, such irrevocable provenance has not yet been discovered for the Pink-headed Duck's painting, except the fact that no earlier painting of the bird is known, from which Latham could have described it. He had certainly not seen a physical specimen of the bird. Thus, for the first time in 1787, Latham used the English name Pink-headed D.[uck],[13] and latinized it in 1790 as *Anas caryophyllacea*[14] or the 'Carnation-pink Duck'.[15,16] In a brief note Latham stated, 'Inhabits various parts of *India*; most frequent in the province of *Oude* (Awadh). Is seldom seen in flocks, for the most part, only two being found together. Is often kept tame.'[17] His source for this information was a 'Mr. Middleton'. Nathaniel Middleton (1750–1807) was a civil servant of the British East India Company, and, briefly,

Resident at the court of Shuja ud-Dowla (1732–1775) of Awadh, Lucknow. He also 'displayed a real interest in Indian art and built up a large collection of . . . natural history drawings by Indian artists.'[18]

The turbulent waters of early nineteenth–century scientific nomenclature tossed *caryophyllacea* between various Anatid genera – *Fuligula caryophyllacea*[19] and *Callichen caryophyllaceum*[20] – but it gradually settled under the evocatively coined genus *Rhodonessa* (rose-tinted),[21] which the German ornithologist Ludwig Reichenbach created in 1853, its binomen becoming *Rhodonessa caryophyllacea*.[22]

The first specimen of the Pink-headed Duck to reach a European museum was a mounted immature male, which Alfred Duvaucel (1793–1824) presented to the Muséum national d'Histoire naturelle, Paris, in June 1825 (though there is some uncertainty over whether the museum acquired it by other means).[23,24] Duvaucel was an intrepid French explorer and naturalist who collected for the museum in South and Southeast Asia from 1818 till his tragic and untimely death in 1824, after a protracted period of ill-health that began after he was tossed by an Indian Rhinoceros in Bihar. He never recovered, suffering additional health issues, sailing ultimately to Madras (now Chennai) in the hope of better treatment, which is where he ultimately succumbed.[25]

The earliest mention of the duck in a British collection was in 1838, when Thomas Campbell Eyton (1809–1880), the sportsman-naturalist noted, 'Few specimens have been brought to this country: we only know of two at present existing; one is in our collection, the other in the British Museum; both were purchased at the sale of the late Col. Cobbe's collection.'[26,27] Colonel Thomas Alexander Cobbe (1788–1836) of the East India Company, hailed from a *fauji* family.[28] He was the political agent to the Governor-General at Murshidabad, Bengal (now West Bengal) from 1831 to 1836, and evidently a keen shikari. After his early death, his collections and trophies were auctioned at Christie's,[29,30] and a 'stuffed' (mounted) specimen of the Pink-headed Duck, from 'N. India', was purchased by the British Museum.[31]

By 1844, when George Robert Gray (1808–1872), the zoologist, author and head of the ornithological section of the British Museum, collated his *List of the Specimens of Birds in the Collection of the British Museum*, now boasting a Pink-headed Duck specimen from Bhutan and Nepal,[32] specimens had begun to trickle into Western museums: Philadelphia, USA (1840), and Berlin, Germany (1843); other 'old' specimens remained undated.[33]

The Pink-headed Duck, ultimately fated with extinction, was truly a recluse. Even a hundred years after Latham's description, its mystique was such that the charismatic Allan Hume (1829–1912) exclaimed in the pages of his catalysing journal *Stray Feathers*, 'There is something odd about this Duck. It must be common somewhere, but *where*, I have failed to discover … Can any one help me to the home of this species?'[34] From that century of trigger-happy collectors and 'sportsmen' of small 'game', who proudly notified their 'bags' [author's emphasis] in extant tabloids or journals, I could glean only ten such records of actual occurrences – through sightings, market purchases or physical specimens. Information about the bird was sparse too. Latham's cryptic notes about its antipathy towards flocking, and docility in captivity were often echoed by subsequent authors.

Pink-headed Ducks were never seen in large rafts or armadas. One would see a brace of them, or if lucky, a paddling, on the still waters of secluded forest pools. They were wary of humans, preferring the quietude of lowland freshwater wetlands in swampy grass jungles, such as 'tanks, pools and nullahs fringed by tall aquatic vegetation, marshy swamps with dense beds of reeds and, in winter, lagoons adjoining larger rivers, regenerated by seasonal flooding. It was not found on rivers or running water of any kind.'[35] Despite printed records showing a wide distribution range in South Asia, mostly from India (especially from the northeastern parts of the country), and a few records from Bangladesh, Bhutan and Nepal, it was indeed a *scarce* bird.

Due to its secretive nature, and rarity, we know very little about its ecology. The birds moved around in pairs or small flocks. They did not mingle with the larger flotillas of ducks but seemed to stay in their vicinity. During the breeding season, April–July, only pairs were seen, or small parties (families?) of birds. They nested in long grass or grass tufts, even away from water, building well-formed circular nests made with dry grass and feathers, measuring about nine inches in diameter and four to five inches deep, in which they laid uniquely (they are unlike any other ducks' egg!) spherical stone-white eggs. Considering this egg, Michael Walters (1942–2017) believed that the bird 'was not closely related to any other duck and possibly represented a relic of an old line that had died out elsewhere.'[36] Closer home, the National Zoological Collection in Kolkata holds five eggs and two skins of the Pink-headed Duck![37] And the Lucknow State Museum holds a mounted specimen of a drake, rather shabby now, but almost relegated to anonymity by an ignorant museum hand's wrong labelling![38] It is the sole survivor of the three that George Reid (fl. 1879–1890) recorded in his catalogues of 1886 and 1890, all procured from the Lucknow area.[39,40] All that we know about their courtship – alas, not from their life of secrecy in the wild, but from the behaviour of captive birds – is that drakes puffed the feathers of their heads and pulled in their heads to rest between their shoulders. Then the neck was stretched upwards and a weak, wheezing call uttered heavenwards. Pink-headed Ducks had a low quack. Both sexes took part in nesting duties. The young fledged in September–October. The birds spread out over a wider area during the cooler weather, November–March, frequenting habitats as mentioned above.[41]

In an age when hunting non-human life was considered a sport and wilderness areas were either combed for fur or feather targets in the dubious pursuit of this frivolous sport, or out of the necessity of providing protein for the pot, the Pink-headed Duck was pursued for its rarity. Rarity is a magnet for both charlatans and connoisseurs. The latter covet it for its lustre and for the fame it will bring, the former for the easy lucre it promises. As far as rare forms of non-human life are concerned, the combination of these two leads to the end of the road for an endangered species. The bird was not a victual delight, except to Thomas C. Jerdon (1811–1872),[42] but it was persecuted for the trade. Collectors and museums were a ready market for unscrupulous trappers and hunters for they paid handsomely. The shy and reclusive hermit of forest pools did not stand a chance against this unrelenting persecution and rapacity.

The last wild bird was shot in 1935, in the Darbhanga District of Bihar.[43] There were several subsequent claims of sightings in the wild, by different people, from different parts of the country, but no one presented a clinching evidence, and hence they were not accepted. Many confounded the Red-crested Pochard with the rare one and foundered. In the opaque mistiness of dawn, crouched amongst tall reeds near still forest pools, the eyes could trick one into agreeing with what the mind wanted to believe!

Julian Hume makes a strong case that collectors escalated the extinction of the Pink-headed Duck.[44] Public and private zoos were popular in the late nineteenth and early twentieth centuries in India, England and the rest of

Europe. Ornamental creatures were their prized possessions, sourced from agents who had global contacts in far-flung continents. It was a trade that paid rich dividends. Despite the heavy demand for the Pink-headed Duck, live specimens trickled in very slowly to the Indian or British markets, a sure sign of its unsocial habits, unlike other waterfowl, and more tellingly, of its rarity. They were trapped in the districts of Goalpara (Assam), and in Purnea (Bihar), in Benoa Chaur (Bihar), in Rangpur (Bangladesh), in Dhubri (Assam) – across the floodplains of the Ganga and the Brahmaputra, in India and Bangladesh. Whatever birds survived the arduous sea voyages (there is no documentation of the number that perished at sea), died in captivity, despite being held in aviaries that had seemingly excellent conditions for their survival. The birds were long-lived, some surviving more than twelve years in captivity, but they did not breed, much to the chagrin of their bird-fancying captors (one of whom exclaimed in frustration that they were very stupid birds because they refused to breed).[45] Snatched from their tropical homes and incarcerated in temperate ones, it was as though the bars of their aviaries, constricting their horizons, switched off freedom's regenerative dynamo. Robbed of a *raison d'etre*, captive drakes displayed listlessly, lifting their fading carnation heads skywards, forcing that procreational wheeze from their peculiar tracheae in a primeval, hardwired display of a hapless, poignantly mortal pantomime. And one by one they died; they died in London Zoo; they died in Lilford Park, Northamptonshire; they died in Calcutta Zoological

Gardens; they died in Berlin Zoo; they died at the Chateau de Cléres, Normandy (itself destroyed by German bombs in WWII); they died in Connecticut, USA; and they perished in Foxwarren Park, Cobham, England. (Their skins became prized possessions in various museums' collections.) Neither did they survive the efforts of the zoöphile Ezra brothers.

Sir David Ezra (1871–1947) was a prominent member of the Baghdadi Jewish community of Calcutta. He was the Sheriff of the city, a director of the Reserve Bank of India, President of the Asiatic Society, and head of multiple industries.[46,47] He also had a large private zoo at his residence where, amongst other creatures, he held Pink-headed Ducks before they could be exported to his avid aviculturist brother, Alfred Ezra, OBE (1872–1955), in England. Alfred also owned a menagerie at his home in Foxwarren Park, Cobham, England. At a time (1923) when 'an unskilled worker averaged Rs 0.5 per day, a skilled worker around double this amount'[48] – Sir David advertised in leading Indian newspapers, his intention to buy Pink-headed Ducks for the royal sum of Rs 100 each! This was indeed a windfall for any prospective trapper, except for one problem – the rarity of the bird. Despite the tens of thousands of ducks shot or trapped and sold in Indian markets, the number of doomed Pink-headed Ducks that were caught or shot was abysmally low. Over the next couple of years, Foxwarren Park received at least fifteen birds. A second reward was never collected! Today, fewer than eighty Pink-headed Duck skins exist in museums. Charles M. Inglis (1870–1974), naturalist and curator of the Bengal

Natural History Society's museum in Darjeeling, had the dubious distinction of seeing the last wild bird in June 1935. The last captive duck died in an Ezra aviary in 1948.[49]

Several subsequent expeditions into remnant Pink-headed Duck habitats in north-eastern India and Myanmar, between 1960 and 2017, failed to locate the bird. The first of these, in 1960,[50] was by none other than 'The First Lady of Indian Ornithology',[51] the sagacious Jamal Ara (1923–1995).[52] The bird eluded others who scoured potential habitats. Rory Nugent[53] did not find it, Jonathan Eames and his team did not find it[54] nor did Richard Thorns and his team.[55] This is not surprising given that India's and Myanmar's burgeoning populations have drained the bird's favoured lowland waterbodies and overrun the exposed lands for agriculture and housing.

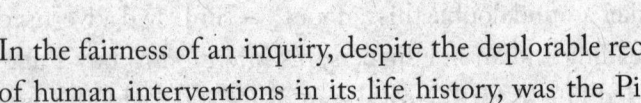

In the fairness of an inquiry, despite the deplorable record of human interventions in its life history, was the Pink-headed Duck at the end of its ecological journey? In an illuminating paper, Ericson et al. observed the following:

> The first molecular phylogenetic hypothesis for the possibly extinct pink-headed duck unambiguously shows that it belongs to the pochard radiation that also includes the genera *Aythya* and *Netta*. It is the sister to all modern-day pochards and belongs to a lineage that branched off from the others more than 2.8 million years ago. *Rhodonessa caryophyllacea* is

believed to never have been common in modern time and we show this has probably been the situation for as long as 100,000 years. Our results suggest that their effective population size varied between 15,000 and 25,000 individuals during the last 150,000 years of the Pleistocene. The reasons behind this are largely unknown as very little is known about the life-history and biology of this species. Presumably it is due to factors related to feeding or to breeding, but we may never know this for sure.[56]

Despite its antiquity, let us recall that the Pink-headed Duck was revealed to Western science through art.

Since that memorable day, nearly 250 years ago, when Bhawani Das picked up his brush, the Pink-headed Duck has been the subject of hundreds of artistic renditions – a *rara avis* aestivating in art. Extant literature is peppered with its likenesses, many of which can now be excavated on the Internet. All are worthy of viewing, for they evoke a strange nostalgia for a diorama now impossible to witness in the flesh. The first illustration of the duck in an Indian work was on Plate 34, in Thomas Jerdon's *Illustrations* in 1846.[57] But I have two other favourite pictures: a glorious plate by A. de Paret depicting a placid jheel, with one drake in the water and three on the near shore, viewed through a gap in the cattails. One of these is shown uttering its wheezing call with an uptilted head, effectively freezing time.[58] And the other, a visceral close-up by Roland Green (1890–1972) of two drakes and a duck, swimming past the viewer.[59]

Black-and-white photographs of the birds in aviaries also exist on the Internet, and one showing ten together is full of pathos. And the reader must also look at a picture of this duck's globular egg.[60, 61]

I first saw a picture of the Pink-headed Duck when it stared back at me from the cover of my first bird book, Sálim Ali's path-breaking *The Book of Indian Birds* (1977).[62] It was in its tenth edition by the time the life-changing mystique of birds gripped me. But in a serendipitous coincidence, the bird had appeared for the first time in the sixth edition of this work,[63] in the year of my birth, 1961 – on a unique plate (No. 64, facing p. 127) – with two of the most endangered birds extant in India besides this duck: the Mountain Quail and the Jerdon's Courser; a fourth illustration was of the common Red-crested Pochard, perhaps placed there deliberately to illustrate the difference in the appearance of the two duck species and dispel any confusion that might arise from their colloquial name, '*Laal Sir*'. The plate was illustrated by the German artist Robert Scholz (1916–1977).[64]

I will close with one last, somewhat bizarre, anecdote regarding a special illustration of the Pink-headed Duck. On 23 November 1994, the postal department of the Government of India issued a set of four multicoloured commemorative postage stamps entitled 'Endangered Water Birds of India'. They had been painted by J.P. Irani, a favourite artist of Dr Sálim Ali. One of these was a

Rs 11 stamp featuring a pair of Pink-headed Ducks. The four stamps were designed, in philatelic terminology, as a se-tenant block of four. These were withdrawn immediately because the Madras Security Printers had negligently printed them with water-soluble inks! Now, whenever they surface in the philatelic markets, these rare stamps are snapped up by collectors at exorbitant prices. This surreal philatelic fiasco is, in a way, reflective of the strange journey of the enigmatic Pink-headed Duck – from the time of its discovery in a painting to its literal fading away from a postage stamp – all within the blink of 150-odd years. All the skins and eggs held in our museums are but sombre reminders of a magnificent dweller on Earth that once existed in a shared space with humanity but lived in its own world.

Any proclamation of extinction rings of a deadly finality, and yet it has been overturned within living memory in India, for at least two avian species – the nocturnal Jerdon's Courser, and the daylight-tolerating Forest Owlet. If '"Hope" is the thing with feathers',[65] then let us dare a similar fate for the Pink-headed Duck and believe that the 'absence of evidence does not mean evidence of absence'.[66]

ACKNOWLEDGEMENTS

I would like to thank Henry Noltie, L. Shyamal, Praveen J., Murray Bruce, Kees Rookmaaker, Rahul Rohitashwa, Haresh and Sita Reddy, all of whom helped with this essay in their own distinct ways.

ABOUT THE AUTHOR

Aasheesh Pittie is a birder, bibliographer and author. He is a co-founder of Indian BIRDS (indianbirds.in) and edited it for its first nineteen years. He is happiest among stacks of books in libraries, or outdoors, experiencing nature. He has created the Bibliography of South Asian Ornithology (southasiaornith.in), a free online database which is also now on the Zotero platform under the same title. His ongoing work, *An Author Bibliography of South Asian Ornithology, 1713–2022* is also online: https://linktr.ee/indiancourser. Besides, he has authored *Birds in Books: Three Hundred Years of South Asian Ornithology: A Bibliography* (Permanent Black, 2010), *The Written Bird: Birds in Books 2* (Indian BIRDS Monograph 4, 2022) and *The Living Air: The Pleasures of Birds and Birdwatching* (Juggernaut, 2023).

When not occupied by the above, Aasheesh can be found immersed in a book, in Indian or Western classical music or in the reality-bending insights of spiritualism.

2

The Rediscovery of the Jerdon's or Double-banded Courser

BHARAT BHUSHAN

The tale of the Jerdon's Courser is surely one of the most dramatic stories in Indian ornithology. Described by Hume in the mid-nineteenth century, the bird remained unseen for decades despite several failed attempts to decipher the puzzle. It was finally rediscovered by the author of this piece, Dr Bharat Bhushan, in 1986. But the re-appearance was short-lived, as the bird has not been encountered since 2008. As a community of scientists and birders, we have failed to find the bird, but as protectors, too, we have failed to take measures that could have sustained the population of this species. Since 2008, there have been many attempts to find the bird, but most have been focused on one place – the Sri Lankamalleswara Wildlife Sanctuary. That may need to change if we hope to, once again, lay eyes on India's most enigmatic courser.

The vivid memory of the relentless rain on 13 and 14 January 1986 will forever be etched in my mind. The thunderstorm roared, hammering away on the tiled roof of the Vontimitta Forest Rest House in Andhra Pradesh, darkening the late afternoon and creating a dusk that should have been at least four hours away. I was worried. Not because of the rain or the thunder, but because I was anxious to return to the forests of Lankamalai hills, near Siddavatam, north of the Pennar River, to search for the Jerdon's Courser (*Rhinoptilus bitorquatus*), a bird that had not been seen or reported for the previous eighty-six years.

The hills of the Eastern Ghats near Vontimitta had swallowed the sun before it had begun its journey towards the western horizon. I could not see the lightning from where I sat on the steps of the Rest House, though the thunder could be heard rolling all around non-stop for more than an hour. I had driven north from Palamaner in the Chittoor District of Andhra Pradesh and arrived at Vontimitta via Tirupati at around 4 p.m., on 12 January 1986. After staying overnight at the Vontimitta Forest Rest House, I was keen to get back on my Enfield Bullet 350cc motorcycle and head, first to Siddavatam and then further north to Badvel, to continue the search for the unseen and elusive bird.

As I sat on the steps thinking that the day was gone, the range forest officer (RFO) of Vontimitta came, wrapped in his enormously huge black raincoat, to tell me that the RFO of the Siddavatam Range had called and informed him that his foresters had permitted Aitanna, a bird trapper, to enter the Lankamalai forests near Reddipalli village, located along the Sagileru River, a tributary of the Pennar. It seemed that he had been 'searching some forest areas on my behalf, for the birds that I had asked him to keep a look out for'.

'Did they mention the name of the bird?' I asked the RFO of Vontimitta.

'Some bird called "*nela nemili*" that Aitanna thinks you would be interested in,' he replied.

I was disappointed as the name *nela nemili* was the Telugu word for the Lesser Florican – a highly improbable sighting or capture for a nocturnal bird trapper in the dense scrub forest area at the foothills of the Lankamalai hill ranges. Before I had travelled south to Palamaner to help Dr N. Sivaganesan, my Bombay Natural History Society (BNHS) colleague, track a herd of seven elephants that had strayed into Andhra Pradesh from Karnataka (that is a different story altogether), Aitanna had volunteered to keep a lookout for the Jerdon's Courser at Lankamalai.

However, I was curious. I knew that Aitanna was aware of the *nela nemili* and its significant difference from the Jerdon's Courser and other cursorial birds (those that do not perch on trees) that are present in the foothill forests of Lankamalai. He had informed me earlier that he had seen the *nela nemili* in the grazing grounds of the Sagileru Valley, towards the Turupukonda hills (of the Velikonda Range). I was sure that he would not place the *nela nemili* at Lankamalai. We had argued about that on an earlier visit. Something was amiss, I felt. Aitanna would not have made such a telling mistake in conveying the information, and he would have also not informed the RFO of Siddavatam unless he felt that it was important to pass on the message to me at Vontimitta. I had to reach Reddipalli, thunderstorm or not, I decided.

Earlier in the day, I had driven north to Cuddapah city and met Divisional Forest Officer Yusuf Sharif of the Indian Forest Service (IFS), to inform him of my presence and my intended survey in the Lankamalai area. He was aware that I was in touch with bird trappers and small game-shikaris from the native communities. To get to Lankamalai now, I would have to cross the Pennar River over a very low causeway on the 50 km road running from Vontimitta to Reddipalli via Bhakarapet and Siddavatam. I decided to risk the drive and reached the causeway over the Pennar, south of Siddavatam town. The water-level markers on the causeway clearly indicated that it was not safe to drive across. I had to return to Vontimitta.

It rained through the night of 13 January and the entire day on 14 January, heavily enough to make day seem like night. As the hours dragged by, I grew impatient. I kept watching the thunderstorm across the Vontimitta Valley, with the Velikonda and the northern Palakonda hill ranges on either side, and the darkened Vontimitta Lake in front foregrounded by the magnificent *gopurams* of the ancient Kodanda Ramaswamy Temple standing in stark silhouette against the rain. This was a tremendous photo montage that I cannot forget, even after all the decades since.

At noon on 14 January, I stopped the Madhavaram-Siddavatam-Badvel bus and requested the bus driver to pass on a message to Aitanna at Reddipalli, about my arrival at Vontimitta and my difficulty in not being able to drive to Reddipalli. The bus driver explained that it might be difficult for him to pass on the message if the causeway was still 'blocked' by the flooded river. The rain slowed down around midnight and the fury abated around 6 a.m. on 15 January. I decided to take advantage of the slow drizzle and rode out of Vontimitta village at 7 a.m.

The overflowing Pennar River and the thunderstorm had damaged the causeway and some of its surface had been washed away. But some motorcyclists were bravely driving over it and I followed, copying them carefully. I eventually made it across the causeway, drove past Siddavatam and reached Aitanna's house at Reddipalli by 8 a.m. His family, his neighbour's family and what seemed like half the village were crowded around the house. They were probably curious about this crazy stranger (me!) who had driven through a thunderstorm and across a flooded causeway just to see a bird.

Aitanna announced that he had trapped a bird that he wanted me to take a look at. He brought it out from the chicken coop where he had secured it safely in a wicker basket. The bird sat comfortably in his closed palms, with its legs dangling between his fingers; it had settled so low in Aitanna's hands that only the top of its head could be seen. At that moment, with all the waiting at Vontimitta, and the continuing downpour at Reddipalli behind me, I was quite tired and waited impatiently for Aitanna to open his palms. I didn't think then that I would get to see the Jerdon's Courser so easily.

Initially, the low neck, the blackish crown and grey upper plumage of the bird made it seem so unlike the Jerdon's Courser. The fleshy, pale-yellow legs had no hallux or 'rear toe', which certainly made it a cursorial species. There was no other characteristic feature that I could relate to. I was about to conclude (to myself) that it had to be some other species when the bird raised its head and stretched its neck. And then the world changed.

Those five to seven seconds, on that fateful morning of 15 January 1986, are forever etched in my mind, stretching for what feels like five to seven hours. I can remember each and every micro-moment of those few seconds. By that very small movement – of raising its head – the quiet, very boring-looking bird settled peacefully in Aitanna's palm, showed off its broad white supercilium, the white throat with the wide rufous band and the narrow, white semi-collar, with the broad brown gorget below. I raised Aitanna's palms to get a better view, and could then see the white collar, margined above and below with a dusky shade below the brown gorget. But it was the faint white mesial line in the centre of the black crown that finally decided it for me. I had kept arguing with myself, during the first five seconds, that it could not be, it could not be, it could not be, and then the 'coronal streak' showed up, as if to decide things. This indeed was the Jerdon's Courser.

Aitanna asked, 'Saar, is this the bird that you are searching for?'

I nodded silently. Aitanna continued, 'Saar, is this the bird that you said nobody had seen for a hundred years?'

I nodded again.

The villagers kept watching this pantomime between me and Aitanna. His eight-year-old daughter stood next to me, looking at the open pages of the *Pictorial Guide to the Birds of India*, showing the colour plate of the coursers and other birds. It suddenly seemed to dawn on her that I had a colour photo–like illustration of the bird in my book, and here was her father, going on about the fact that nobody had seen the bird for a hundred years. She interrupted the two of us, and exclaimed, 'How can you say so? See, this Saar has the bird in his book. How did he draw the bird if he had not seen it?'

The innocence of Aitanna's daughter brought me back to reality. I tried to answer her, 'Not at all. This is just a painting, drawn from a dead bird. Nobody has seen this bird alive for nearly a hundred years. Your father has found it, alive, and I am going to announce it to the world. This is an important bird.'

And the young girl replied, 'But why? Why did my father catch it to show it to you? Did you come all the way from "Bambayee" only to see this bird? Why do you want to see a bird if nobody has seen it for a hundred years?'

Why, indeed? Why should one be interested in a bird if it had not been seen for nearly a hundred years? Why should one travel across the country in search of that bird? What was so important about the Jerdon's Courser that BNHS had sent me specifically on this survey?

How could I explain the significance of the Jerdon's Courser to the young girl? Would the villagers understand that their village would now be on the world map? Would they believe me if I told them that Aitanna would now be famous? How could I explain that this small elusive bird would do all that? That this bird would change their lives? And mine.

As part of the preliminary survey, I had been visiting Reddipalli over the previous six months and had met Aitanna and other bird trappers and small game-shikaris from various communities in the region. The village was on the Siddavatam–Badvel section of the Cuddapah–Nellore highway in southern Andhra Pradesh. Both Vontimitta and Siddavatam were very striking *mofussils* with two very prominent and ancient temples, revered through the ages in the cultural history of the Andhra region. Nestled in the Eastern Ghats, amidst different hill ranges, these two largish villages or semi-towns were ideal field survey locations for they allowed easy access (and movement) to the numerous valleys and gorges of the hills. The Pennar and Sagileru Rivers provided the East–West and North–South divides and had helped plan the surveys for the Jerdon's Courser since June 1985.

In the early 1980s, the United States Fish and Wildlife Service (USFWS) funded a research project of the BNHS called the 'Study of the Ecology of Rare and Endangered Species of Wildlife and Their Habitat'. Under this project, I was stationed at Karera, in the Shivpuri District of Madhya

Pradesh, to study the rare and endangered Great Indian Bustard (GIB). Later, upon repeated sightings of the GIB in Andhra Pradesh, in March 1985, along with another BNHS colleague, I travelled to Rollapadu in the Kurnool District of the state to set up a field research station for the study of the bustard. In April 1985, Mr J.C. Daniel, the Director of BNHS, and my mentor and guide, asked me to accompany him to the Kurnool and Chittoor districts to survey two forest areas and develop proposals for conservation under the Government of India's 'National Wildlife Action Plan'. And yet again, that is another story.

After the survey at Chittoor, Mr Daniel was to travel southwards, to BNHS field study locations at Vedanthangal and Point Calimere (both in Tamil Nadu), leaving me behind at Tirupati with a jeep and a driver. This was indeed my opportunity, I thought. I requested Mr Daniel to permit me to survey southern Andhra Pradesh during the month that he would be away at Point Calimere. Specifically, I wanted to survey the Pennar river valley areas in southern Andhra Pradesh and try to establish whether the Jerdon's or Double-banded Courser was present or actually extinct in the region. At that time, the previous sighting had been nearly eighty-six years ago, and the species was known only from the Pennar river valley area in southern Andhra Pradesh and near Bhadrachalam, along the Godavari river valley, in the northern districts of the state.

A preliminary survey was conducted during June–July 1985 along the Pennar river course and adjoins from Penakacherla in Anantapur District to Jammalamadugu in

Cuddapah District, within the Siddavatam and Vontimitta areas (also in Cuddapah), along with Somasila and Rapur in Nellore District, and the Nagari hills of Chittoor District. Days between field surveys were used to meet local forest officers and explain to them the reason and importance of the survey in their areas. Based on the information collected from local sources during the preliminary survey, spot-survey locations were decided upon and surveys were conducted at Somasila in Nellore District and Siddavatam in Cuddapah District between July and August 1985.[1] This groundwork led to the eventual rediscovery of the Jerdon's Courser in January 1986 at Reddipalli, in the Lankamalai hill ranges of Cuddapah District.

Sadly though, since 1986, in spite of the intensive field studies by subsequent scientists, appropriate field research (using radio telemetry) is yet to begin. The species is almost impossible to understand without proximate and accurate knowledge. Radio telemetry and the use of satellite-linked telemetry will help track individual birds or pairs and determine their breeding locations and whether they are locally resident or transient (migrating to other locations). The significance of the rediscovery, the endemic status of the species, the confined distribution, the rarity of its numbers and the challenge (yet to be surmounted) in studying it require that many diverse perspectives be documented and presented appropriately.

Even today, the species continues to humble the efforts of science to unravel its mysteries. How much more is left to understand about the Double-banded Courser? Where else could it be? How many survive? Are they safe? Are they about to become extinct? With so many questions and so few answers, historical knowledge and perspectives about the species need to be put together, and perhaps this is the best time to do so.

Dr Thomas Claverhill Jerdon (1811–1872), a British physician-zoologist-botanist of the East India Company, stationed with the Madras Regiment as the surgeon-major, first 'procured' the Double-banded Courser in c.1848 and sent the specimens to Edward Blyth (1810–1873), the curator at the Museum of the Royal Asiatic Society of Bengal in Calcutta, to study and determine if they were to be recognized as a new species. Jerdon, in his important publication *The Birds of India* recorded the species 'from the hilly country above the Eastern Ghats, off Nellore and in Cuddapah'.[2] Jerdon was a prolific naturalist and writer who had extensively surveyed the Eastern Ghats of the erstwhile Madras State (now southern Andhra Pradesh). On the basis of his bird surveys, he wrote in 1877 that he believed the Double-banded Courser to be a 'permanent resident' and an 'almost unique instance of a species of Plover having such an extremely limited geographical distribution'.[3] This emphasis was quite intriguing, and became an important perspective of my surveys more than a century later.

A later edition of Jerdon's *The Birds of India* (Volume 1) includes a short memoir written by W. Elliot in 1873.

Elliot, while listing the former's army postings, states that Jerdon had '... an opportunity of seeing a part of the country difficult to access and rarely visited; and he did not neglect it, as his notices of the birds of the Eastern Ghats subsequently showed.'[4]

Elliot further notes, 'After passing about four years with his regiment, he obtained leave of absence to visit the Nilgiri hills, where he was married in July 1841. Six months afterwards, he was appointed Civil Surgeon of Nellore ...'. This brought Jerdon to the peneplains (lands that lie between a hill range and a coast) along the eastern coast, looking westwards to the Ghats. The memoir also mentions that 'the wilder parts of this country between Madras and Nellore are occupied by the Yaanadis, a remarkable aboriginal tribe, of semi-nomad habits, subsisting on the spontaneous produce of the jungles, and possessing in consequence a minute acquaintance with the forms of animal and vegetable life around them. By their means, Dr Jerdon discovered many new species ...'

Between 1862 and 1864, Jerdon wrote that he believed the Double-banded Courser to be 'spread through many parts of Balaghat (south-central Madhya Pradesh) and Mysore (Karnataka)'.[5] This could particularly refer to the Bastar region of Balaghat, and the Anantapur areas of the erstwhile Mysore State. It is entirely possible that – similar to the assumption that the species could occur between Bhadrachalam and Cuddapah – Jerdon presumed that it was also distributed in regions adjoining these two locations. The plains-forests below the hilly regions – the habitat of the

Jerdon's Courser – continue from Sirroncha (northwards, along the Indravati River) through to the Bastar and Kanker districts of Chhattisgarh (erstwhile Madhya Pradesh and historical Central Provinces and Berar State and Bastar Kingdom).

Blyth, to whom the type specimen was sent, named it *Macrotarsius bitorquatus*, ascribing it to a separate genus *Macrotarsius*, making it distinct from the other coursers known from the Indian subcontinent (grouped under the genus *Cursorius*), and credited the discovery to Jerdon in the *Journal of the Asiatic Society of Bengal*, in two publications – 1848[6] and 1849.[7] The species name, *bitorquatus*, means 'double-banded' or 'two coloured-plumage rings around the neck'. He located the type specimen to be from the 'Eastern Ghats of the peninsula of India'.[8]

In 1852, the English geologist, ornithologist, naturalist and systematist H.E. Strickland placed the *bitorquatus* with other congenerics (birds of the same genus as *bitorquatus*), under *Rhinoptilus*; this shift was recognized by the zoologist R.B. Sharpe in 1896,[9] and subsequently by other ornithologists. In 1929, Stuart Baker followed Sharpe in assigning *bitorquatus* to the genus *Rhinoptilus*, as did J.L. Peters in 1934 and Dillon Ripley in 1952 (though the latter erred later, as we shall see). Ripley agreed with previous arguments made by British geologist and naturalist William Thomas Blanford that the genus *Macrotarsius* was a 'very weak one',[10] and pointed out that the *bitorquatus* with the 'darker patch on the breast approaching a breast band'[11] seemed to be more common to the *Rhinoptilus* species.

Though Ripley did point out the variance from *Cursorius*, he seemed to have erred at that time in continuing, in his *Synopsis of the Birds of India and Pakistan* (1961),[12] to list *bitorquatus* in the *Cursorius* genus. There was no logical reason for the genus to be moved to *Cursorius*. It was just placed there, in the literature of birds in India, without any perception, perhaps because it was a courser. In the *Handbook of the Birds of India and Pakistan* as well, which Ripley wrote along with Sálim Ali between 1964 and 1974,[13] all Indian courser species were placed in the genus *Cursorius*. Similarly, other scientists – perhaps wrongly – considered *Rhinoptilus* as not sufficiently distinct from *Cursorius* in terms of generic status and treated all of the African forms as *Cursorius*.

In this manner, the Jerdon's or Double-banded Courser came to be included as *Cursorius bitorquatus* in the checklists for Indian birds, though it was obvious that it should be listed within *Rhinoptilus*. All *Rhinoptilus* congenerics, including the three in Africa, are crepuscular and mostly nocturnal. This is important, for understanding the species to be of the genus *Rhinoptilus* helped in its rediscovery in 1986, which would not have been possible if one were to search for a *Cursorius* congeneric.

To return to the record of sightings of the species, William Thomas Blanford recorded the Double-banded Courser in May 1867, near Sirroncha in Maharashtra, and in March 1871 at Bhadrachalam, near the Godavari river valley in northern Andhra Pradesh.[14] The first sighting was at a point 24 km east of Sirroncha, near the Godavari River, which would place the location near the southern banks of the Godavari, after its confluence with the Indravati River.

The last 'presumably authentic sighting'[15] was by Howard Campbell in 1900, in the Pennar river valley near Anantapur. Subsequent surveys in the region, including the Ornithological Survey of the Eastern Ghats by Hugh Whistler and Norman Boyd Kinnear in 1930, and the Hyderabad State Ornithological Survey by Sálim Ali in 1931–1932, failed to locate the Double-banded Courser. Similar failure by subsequent surveys led to the Double-banded Courser being considered as one of the rarest avian species in the world. My hypothesis is that the surveys were looking for a *Cursorius* congeneric in the 'river-plains' of the river valleys, while they should have been looking for a *Rhinoptilus* congeneric in the foothill-forests and scrub of the hills. Jerdon had not mentioned the 'river valley' in his description; instead, he had specified that the bird had been procured 'from the hilly country above the Eastern Ghats'.

Based on the lack of further records and sightings after Blanford in 1871 and Campbell in 1900, Sálim Ali and S. Dillon Ripley stated in their *Handbook* that the Jerdon's or Double-banded Courser was known only 'from a restricted area in eastern India, from the valley of the Godavari River, near Sirroncha and Bhadrachalam, and from Nellore, Cuddapah and Anantapur, in the valley of the Pennar River'.[16]

More than 40 years after Sálim Ali's Hyderabad State Ornithological Survey, BNHS mounted two 'special explorations'[17] in 1975 and 1976, that were conducted in collaboration with the Smithsonian Institution (based in the United States) and the World Wildlife Fund, India. The surveys were supported by a colour poster prepared by J.P. Irani, showing the Jerdon's Courser with the Indian Courser. The sketches showed both birds (the *Cursorius* and the *Rhinoptilus*) in similar postures, which I believe could have been misleading. The poster, which was circulated in Andhra Pradesh and adjoining states, did not receive any response.

Failure to locate the Double-banded Courser resulted in the species being listed as extinct or presumed to be as such. Sálim Ali and S. Dillon Ripley, in their 1985 proposal (unpublished) to the Government of India to conduct 'Distribution and Ecological Studies of Relictual Avian Populations on the Indian Peninsula' by the Smithsonian Institution, wrote that the effort to search for the Jerdon's or Double-banded Courser had been 'too little'. Though it was the opinion of two of the most prominent ornithologists of our times, it was highly improbable that a species that was never threatened by hunting or never known to live in a critically fragile habitat, but reported (and known) from a vast region, with potentially suitable existing habitat, could have become extinct.

The fact that the Jerdon's Courser was truly a peninsular endemic and extremely rare due to the paucity of records, only served to make the species exciting enough to be considered a rarity worth seeking. It seemed highly improbable that the species would be extinct given that its reported range extended from Sirroncha to Bhadrachalam in the Godavari River region and southwards to Anantapur, Cuddapah and Nellore in the Pennar River region. Jerdon had found the Double-banded Courser to inhabit 'rocky and undulating ground with thin forest jungle' and believed that the species was a 'mountain form of *Cursorius* (Indian Courser), frequenting rocky hills with thin jungle'. Blanford also wrote that he recorded the species in 'thin forest or high scrub, never in open ground', emphasizing that he 'never saw any on hills'. Ali and Ripley also specifically wrote in their *Handbook* that unlike the Indian Courser, the Jerdon's Courser was not found in open wasteland.

The specimens collected by Jerdon and later by Blanford, did help in determining certain vital field characters. Firstly, the sexes were alike. The tail was clearly white and black, and a white wing-bar was prominent in flight. Jerdon described the Double-banded Courser as 'found in small parties, not very noisy, but occasionally uttering a plaintive cry,'[18] while Blanford met 'pairs twice and three birds together, once in 1867 and 1871'.[19] He also described the flight as more rapid than that of the Indian Courser.

All of these proved to be useful pointers in the 1986 rediscovery, which validated the search for a *Rhinoptilus* congeneric instead of a *Cursorius*. Following the rediscovery and my documented insistence that the Double-banded Courser was an obvious *Rhinoptilus*, Ripley and B.M. Beehler presented in 1989 the 'results of a systematic analysis of Jerdon's Courser and its nearest allies and discussed the taxonomic status of the genera *Rhinoptilus* and *Cursorius*'. They studied and compared the Jerdon's Courser to eight other birds of the same family from India and Africa. Based on their analysis, they concluded that the Double-banded Courser was more closely related to the African coursers and clearly belonged to the genus *Rhinoptilus*.[20] Ripley and Beehler concluded that the *Rhinoptilus* is defined by the expression of three unique characteristics: throat characters, such as the posterior dark band, the mottled middle band and the behaviour of nocturnal activity. They further determined that the peculiar throat-banding and predominantly nocturnal habits are characters specific to the genus, 'unambiguously distinguishing this clade from any members of the genus *Cursorius*'.[21] This had certainly been made clear by the manner of its rediscovery in 1986 and my subsequent sightings of the Double-banded Courser in the foothill forests of the Lankamalai hill ranges.

It is interesting to note that in the years since its rediscovery – at this time of writing in February 2024 – the nest of the Jerdon's or Double-banded Courser is yet to be located in the field. The *Handbook*, however, informs that in 1895 the defunct *Asian* newspaper published an anonymous note describing the nest with a 'clutch of 2 eggs, bright yellow-stone; the ground colour almost obliterated by black scrawly blotches-and-spots; (and) laid on the ground in thin scrub jungle.' The *Handbook* also describes that the two males collected by Blanford on 5 and 8 March 1871, near Bhadrachalam, showed no gonadal development. These specimens are kept in the British Museum of Natural History at Tring, UK.

The literature discussed above helped me plan the 1985–86 search and survey – Jerdon had mentioned the 'hilly country above the Eastern Ghats, off Nellore, and in Cuddapah', and Howard Campbell had mentioned the Pennar river valley areas near Anantapur. These areas comprised the Rayalaseema region in southern Andhra Pradesh and included the hill ranges of Seshachalam, Palakonda, Erramalai and Velikonda of the Eastern Ghats. It also broadly includes the Nagari hill tracts south of Nellore and Chittoor, and the Nallamalai hill ranges to the north of Cuddapah, bordering Kurnool.[22] Though the word 'Rayalaseema' is taken to indicate the regions in what was the southern border of Emperor Krishnadevaraya's kingdom, in the colloquial and contemporary times of the 1980s, it was also meant to refer to the stony scrub areas that formed the edge of the Deccan Plateau as it met the Eastern Ghats.

Mr J.C. Daniel had advised that I limit my travels to the southern districts of Andhra Pradesh, and leave Bhadrachalam for a later survey. The Pennar River, a prominent feature of the Rayalaseema area, flows through Anantapur, Cuddapah and Nellore districts (recently bifurcated into two districts each) before meeting the Bay of Bengal. Thus, the river course was considered as the first point of reference for the surveys in southern Andhra Pradesh.

The next point of reference was to limit the survey to certain types of habitats. Sightings of the Double-banded Courser had been reported from 'rocky and undulating ground with thin forest jungle'. Jerdon had, however, made a very specific mention – he believed the bird to be a 'mountain form of *Cursorius*, frequenting rocky hills with thin jungle'.[23] Blanford added to this specific habitat type, mentioning that it was found 'in thin forest or high scrub, never in open ground,'[24] and concluded that he 'never saw any on hills'. Initially, I thought that he was contradicting Jerdon, who had mentioned 'rocky hills'. Later, during the survey, while speaking with local bird trappers and foresters, I realized that Jerdon and Blanford had agreed with each other. Finding 'rocky hills with thin jungle' and 'thin forest or high scrub' did not require one to climb hills or go out on open ground, but simply implied *thin and open scrub near rocky hills*. The bird would not be found in the foothill forests below evergreen forests, or within forests on hill slopes, or on open grazing grounds, and I would not have to search near any wetland.

This was a bird with a rather narrow and specific habitat type, almost with its own niche. Moreover, the Double-banded Courser would not be inhabiting areas favoured usually by the Indian Courser. The BNHS poster with the colour painting by J.P. Irani, showing both the Jerdon's and Indian Courser, was once again distributed, but this time along with a descriptive note in Telugu, mentioning that the habitat could be different, and the bird could be found in the hilly areas. A hundred copies were distributed personally, in Anantapur, Cuddapah, Chittoor and Nellore districts, to select individuals, such as foresters, local shikaris, bird trappers and native forest-dependent communities.

The chosen 'informants' were later asked to identify local birds from the pages of the *Pictorial Guide to Birds of the Indian Subcontinent*, written by Sálim Ali and Dillon Ripley. This helped me to verify the local informants' extent of knowledge on birds, their habitats and their local names in Telugu or Deccani Urdu. These enquiries were followed up with field surveys in the area, along with the local informants, to determine if it indeed was the possible habitat type for the Double-banded Courser. Through the process of elimination, I was able to locate three bird trappers, who gave three different versions of what could have been the Double-banded Courser: two in the Siddavatam area in Cuddapah District, and one in the Penakacherla area in Anantapur District.

The vast data that was collected through this process of the names of various birds in Telugu delivered an added bonus. I got married in December 1986, and my wife Thulasi, a student of linguistics, was planning to study classical literature for her PhD research. This would have meant staying back in Chennai, to read voluminous books in a dimly lit library hall, while I would be moving about, supposedly on an adventure (according to her), in the Eastern Ghats. So, we decided to converge ornithology and linguistics, by changing her PhD study to the 'Linguistic Analysis of Bird Names in Telugu', which was completed at the Department of Linguistics, University of Mumbai. This helped us travel together in the Rayalaseema areas, from 1986 to 1989.

Lankamalai. The word itself defines the hill ranges. For it indicates an isolated hill range, not contiguous with the other hills of the Eastern Ghats. It rises up from the south-eastern edge of the Deccan, before the peneplains and river valleys split the area from the northern Palakonda and Velikonda hill ranges. The hills range from the northern banks of the Pennar River, towards the Nallamalais at Ahobilam and Nandyal. The Palakonda hills are separated from Lankamalai at Kanamalopalle and Vontimitta, along the southern banks of the Pennar. The north-to-south Sagileru River flows through the valley, between the Lankamalai and the Turupukonda-Velikonda hills, before the confluence with Pennar.

The Siddavatam area is about 20 km from Cuddapah city. In 1986, barely 11 km outside Cuddapah (towards Siddavatam), and all around the foothills from Reddipalli to Badvel, one could see gentle slopes with open scrub forests on the foothills. This habitat is true of the western slopes of the Velikondas, which face the Lankamalai, and also seemingly perfect in the lower slopes of the Turupukonda hills, which is closer to a saddle ridge that connects both these ranges and rises higher than the Sagileru river.

With the help of the RFO of Siddavatam and the Kumbagiri forester, I was introduced to Aitanna, a resident of Reddipalli village. The village was nestled below the saddle ridge and had a favourable location, with open scrub foothill forests in the west, south and east. Aitanna was a *chinna vetagadu*, that is, a small-time or small-game shikari. Surprisingly, along with three other villagers, he recognized the Lesser Florican, known locally as *nela nemili*, from the pages of the *Pictorial Guide*. That made me suspicious of his claims, as the habitat closer to the village seemed highly unsuitable for the florican. We had an argument before I realized that the local bird trappers were testing me. To my question, Aitanna laughed and pointed at the grazing grounds towards Turupukonda, clarifying that the florican was only seen after the first rains, when the grass is taller than the village dogs. He could not, however, identify or speak about the Double-banded Courser from the poster that I had circulated, or from the pages of the *Pictorial Guide*.

As an afterthought, Aitanna spoke of Pichchanna, an experienced bird trapper and a Yaanadi shikari, known to be an expert on small game and birds. He also informed me that the 'legend' was settled as a custodian of sweet lime and mango orchards near Kumbagiri village, about 12 km north of Reddipalli. The local forester was also in agreement and spoke almost in awe of Pichchanna's skill in the forest. According to him, it was impossible to catch the veteran while he was hunting or find any evidence of him moving about in the Lankamalai forests.

We first met Pichchanna in the presence of the forester, and he was therefore possibly reticent in my first discussion with him. He recognized the Double-banded Courser from the poster with some reluctance and hesitation. He said that it could be what was locally known as *Kalivi-kodi*. On hearing this, Aitanna and his fellow bird trappers, opened up the folded BNHS posters and began to re-examine the details. They began to nod at one another, agreeing with

Pichchanna, and pointed out that the BNHS poster was misleading. Otherwise, they said, they would have identified the bird as the *Kalivi-kodi* rather promptly.

The word *Kalivi* is the Telugu word for *Carissa*, which is a common scrub vegetation along with *Zizyphus* and *Acacia* in the valley. Pichchanna explained that the name *Kalivi-kodi* was the best description for the bird's habit, habitat and its behaviour while hiding amidst the thorny *Carissa* bushes. He had seen the bird only during the rainy season in the foothill forests, possibly because he moved his snares and partridge and quail traps into the hills during those months. At other times, when it was drier in the foothills, Pichchanna would place his bird traps in the lower valleys, nearer to the water streams and drains, at the edge of the forests.

Of course, during the entire discussion, Pichchanna kept stating vehemently that he had stopped shikar 'many years ago'. This could have been because of the presence of the Kumbagiri forester. In further discussions, Pichchanna provided the clincher. He described the Red-wattled Lapwing as the *utha-titti* and *uththithi*, explaining that the bird is known in the nearby Chittapalli village as the *uththutha-gaadu* – meaning 'stammerer'. The name *uththithi* for the Red-wattled Lapwing was recorded in my earlier surveys, all across the Rayalaseema districts, but the exact phonetic, *utha-titti*, had only been mentioned here, by Pichchanna.

Utha-titti was, of course, a phonetic corollary to the *Adavi-wuta-titti* in the 'language of the natives', mentioned in 1877 by Jerdon as being the Telugu name of the Double-banded Courser. The 'natives' were probably the Yaanadis of the Nellore, Cuddapah and Chittoor areas of Andhra Pradesh. Jerdon was known to have depended extensively on the Yaanadis of Nellore hills in the Eastern Ghats for his field collections of birds, insects and mammals.

I was curious. How did Pichchanna, a Yaanadi on the western slopes of the Velikondas, know the exact phonetic, *utha-titti*, while the locals described the Red-wattled Lapwing as *uththithi*? Pichchanna then explained that he used to live in a village between Kaluvoya and Somasila, in Nellore District, on the eastern slopes of the Velikondas, but was now settled at Kumbagiri. It made perfect sense, I thought, for Jerdon had indeed picked up the phonetic *wuta-titti* while on the eastern slopes.

The name *Adavi-wuta-titti* was possibly translated for Jerdon by non-Telugu speakers, in a word-by-word rendition, as 'Jungle-empty-purse'. In the local Telugu communities, women carried coins in a cloth-lined purse, called *tiththi* or *titti*. This would have never referred to the name of any bird, but would have been explained as such, if the three separate names (*Adavi-wuta-titti*) were translated for the European Sahib by the subaltern or a local native assistant. The key of course was the first word 'Adavi', which translates as forest or jungle. It was a revelation, that day, to understand that the Double-banded Courser had been thought to be a 'forest lapwing'. Its very name pointed to the fact that a birder should not be searching for the species in the habitat of the Indian Courser.

Meanwhile, some months after our first meeting, in October 1985, the forest department employed Pichchanna at the Kumbagiri forest nursery compound. When we met him subsequently, he explained, hesitantly, that he could not possibly go into the forests now to place snares and traps to catch small game or birds. He said that he could, however, guide others to search for the Double-banded Courser in certain locations. Aitanna was very enthusiastic and wanted to go ahead, explaining that he had been keeping a watch on the foothill scrub forests since my earlier visits to the area. He even claimed that he had sighted a few birds at night, when searching with powerful torches. I was quite skeptical of this claim until he explained his method, which reminded me of the 'mirshikaris' from Bihar's Kaabar Taal, who were bird trappers working with BNHS at the Keoladeo Ghana National Park in Bharatpur.

This is how the search method worked: during his nocturnal searches, Aitanna used an 18 inch long, four-cell torch, which had a powerful beam, along with a tremendously annoying and extra-loud electric buzzer, connected to a 350cc Bullet motorcycle battery, which he hung over his shoulder, over a double-layered rubber sheet salvaged from a motorcycle tyre tube. The rubber sheet would prevent any injuries or burns if the hot battery fluid spilt over onto his chest. The utility of the electric buzzer was its extremely high-volume continual sound, which would hide his footsteps or any other human presence, as the *shikari* walked through the thorny scrub forests. Aitanna would point the torch ahead to check for the presence of small game, and would then immediately swing it away to create an abnormal darkness in the eyes of any creature that happened to be around. Most small game would freeze, unsure of what had happened. Aitanna and the other local trappers, as they later demonstrated to me, would then just rush up to pick up the bird or animal. In this manner, they would catch grey junglefowl, partridges, quail, hare and, on occasion, even a four-horned antelope.

Taking the local reports into consideration and comparing them with the habitat niche and behaviour of the *Rhinoptilus* congenerics in Africa, a search pattern was developed sometime in December 1985. The essence of this was that the Double-banded Courser was possibly found in thorn scrub forests, and could be a ground nesting bird, which does not build any nest; it probably spent the day under the shade of a thorny scrub bush, and was thus able to evade observation; in effect, it was a crepuscular and/or nocturnal species.

It was this stratagem, distilled after multiple rounds of surveys, meetings with local people and trappers, that led to that rainy morning of January 1986, when the *Rhinoptilus bitorquatus* finally revealed itself.

Where can the Double-banded Courser be found now? The Jerdon's Courser has historically been known to be endemic to southern India, especially in erstwhile undivided Andhra Pradesh, except for Sirroncha, which lies on the border between undivided Andhra Pradesh and neighbouring Maharashtra. Although Bhadrachalam – another potential location – is now within the geographical area of Telangana State, for current purposes we will consider the undivided state of Andhra Pradesh to define the distribution of the bird. Considering the distance between Cuddapah and Bhadrachalam, nearly 700 km, and the continuity of similar habitat between these locations, it would be highly improbable that the Jerdon's Courser is not present here in an appropriate niche. The reason that the species is not known from these regions nowadays is simply because nobody has searched with diligence and persistence.

The Sirroncha area is located 24 km east towards the Godavari River. I have surveyed these areas and if the species is present at all, it would have to be in the region between Pankana and Burugudem in Karimnagar District, and in the regions of Chandrupatla, Peruru, Chikupalle, Pamunuru and Edicharlapalli, along the Godavari River in the Khammam District of Andhra Pradesh.

The mention of Mysore by Jerdon could be misleading as it would probably allude to the Anantapur record, which may have been part of the erstwhile Mysore State. An unconfirmed and usually less-discussed record is of the species being 'listed without comment' from the immediate vicinity of Madras (now Chennai), Tamil Nadu, by Douglas Dewar in 1905. This could probably be the south-eastern extremity of the appropriate habitat of the Jerdon's Courser – in the Narayanavanam and Nagalapuram hill ranges of the Nagari Hills in the Eastern Ghats. I would predict that if at all the Jerdon's Courser is commonly found in any area, it would be the Nagari Hills. These regions are mostly undisturbed, with areas of contiguous plains-forests below hilly regions.

Since the rediscovery, there have been many sightings of the Jerdon's Courser by expert birders, sometimes with Aitanna's help, in the Reddipalli scrub areas of the Lankamalai hills in the Eastern Ghats hill ranges of southern Andhra Pradesh, from 1986 onwards. Again, the species has not been reported or sighted elsewhere in its known range, because of the lack of appropriately methodical surveys. In the years following 1986, field scientists from BNHS have continued to attempt studies on the species – and some have even managed to record the call of the bird – but they are yet to unravel the puzzle of the Jerdon's or Double-banded Courser. That is what makes the bird so interesting and special. This is a species that has always been there, within a narrow foothill forest stretch. The Jerdon's Courser has a tremendously confined niche, which should make it easy to spot and study. However, this is not so. It has amazing crypsis, specific time slots for movement within its habitat, selective foraging habits and the most perfect nesting behaviour which renders the bird almost invisible. A nest is yet to be located in the natural habitat of the bird. An egg collected at Kolar in 1917 by Ernest Gilbert Meaton, now at the Zoology Museum of the University of Aberdeen, was later decoded by the museum's curator, Alan Knox, to be that of the Jerdon's Courser. That, probably, is the only known egg of the species.

So, is it possible to find the Jerdon's Courser again? At present, nearly 160 years since its discovery by Jerdon, the Double-banded Courser is known only from the Lankamalai, Velikonda and Palakonda hills of the Eastern Ghats. During my surveys in the region – from 1985 to 1992 and later during short visits in 1993–1995 – I have confirmed reports and records of the Jerdon's Courser in various locations in the hills of the Eastern Ghats in southern Andhra Pradesh. These locations include the Lankamalai and Turupukonda foothills, and Velikonda, Chitvel, Sanipaya and Rayachoti scrub forests.[25]

It will take a systematic search of these habitats and other suitable areas to make contact with India's most enigmatic cursorial species once again.

ABOUT THE AUTHOR

An MSc and PhD in Field Ornithology, BNHS, University of Mumbai, Bharat Bhushan rediscovered the Jerdon's or Double-banded Courser, in January 1986. He was also part of the team that rediscovered the endemic Golden Gecko at Talakona in the Eastern Ghats, near Tirupati. His doctoral studies on the birds of the Eastern Ghats in southern Andhra Pradesh were carried out in splendid forests, which have a religious significance, in the Seshachalam, Palakonda, Velikonda, Lankamalai, Erramalai and Nallamalai hill ranges. He was the field scientist and conservation officer, BNHS (1982–1992), on the Endangered Species Project and the Bird Migration Project and, later, professor and dean (Academics) at the Yashwantrao Chavan Academy of Development Administration (1996–2022), Government of Maharashtra. He is currently an adjunct professor at the University of Mumbai, and is also the honorary secretary of BNHS. Bharat Bhushan has authored the book *Birds of Ramayana*.

3

A Tale of an Absconding Owl

PAMELA C. RASMUSSEN

Can an owl hide in plain sight, especially one that is active during the day? Apparently so, if the humans in search of it have been completely misdirected about its location. For years, birders and ornithologists, including Dr Sálim Ali and Humayun Abdulali, looked for the Forest Owlet in several locations, from Melghat in Maharashtra to Mandvi in Gujarat. Since the time Dr Pamela C. Rasmussen found it in 1997, the owl has turned up in multiple locations; so while the story of the rediscovery criss-crosses the world, the owl is now known from the Tansa Wildlife Sanctuary, right in Mumbai's backyard. This essay is the story of that rediscovery, in the discoverer's own words. A brilliant example of fieldwork backed by painstaking study of museum specimens from across the world, this tale will tell you what it takes to find a near-mythical species.

For a birder wishing to see the Forest Owlet (*Athene blewitti*), it is now an eminently achievable goal. In fact, with some planning and effort, it can be realized in a day trip from Mumbai to the Tansa Wildlife Sanctuary, whose eBird bar charts show the species to be one of those that is most frequently encountered. But this was not always so.

Soon after beginning work at the Smithsonian Institution as scientific assistant to Secretary Emeritus Dr S. Dillon Ripley in the early 1990s, I became acutely aware of the 'mystery birds' of India. These included, of course, the Pink-headed Duck, Jerdon's Courser (which had just been rediscovered) and the Mountain Quail, as well as the Forest Owlet. Both Ripley and Dr Sálim Ali had long-standing interests in all these species[1,2] (they and S.A. Hussain searched for the Forest Owlet in Melghat, Maharashtra and Mandvi, Gujarat, but to no avail). Ali even referred to the Forest Owlet as an 'absconder'! Subsequently, Ripley and Ali began to have doubts not only about the survival of the species, but whether it was even a species, and even if so, whether it could be positively identified in the field.[3,4] I, too got mildly interested in the Forest Owlet, believing that further searches along the Satpura Range might pay off, but I wasn't able to follow through on these plans. Numerous authors had written it off as extinct, while others were more circumspect in calling it possibly or probably extinct.[5]

A few years later, alerted to problems flagged by British ornithologist Alan Knox regarding the Meinertzhagen collection,[6] Dr Robert Prys-Jones and I began studying the specimens at The Natural History Museum, UK (NHM), that formed the basis for Meinertzhagen's important Indian subcontinent bird records, with several of them being the only evidence for a particular taxon in the region. What we found within a short time was that nearly all these singular records by Meinertzhagen were fraudulent – he had simply stolen most of them, relabelled them with false data as if he were the collector and published them with often detailed but fanciful accounts of his encounters with them.[7,8]

Shortly after returning from this first museum visit that was largely concerned with examining the veracity of the Meinertzhagen South Asian records, I was sitting in the office when Dr Storrs Olson happened by and asked me something about the Forest Owlet. I picked up the nearest copy of Ali and Ripley's *Compact Handbook*[9] and started reading aloud, my voice rising abruptly and involuntarily when I read 'the latest [specimen] in October 1914 at Mandvi on Tapti River . . . by Meinertzhagen'. I was truly taken aback, and quite suddenly I just had to know more. One question led to another. If Meinertzhagen was the last authority for the existence of the Forest Owlet, surely that record was suspect; so when and where was the last valid record? Could it be proven that Meinertzhagen's *blewitti* specimen was fraudulent? If his Forest Owlet specimen had led to a search in the wrong locality, had other sites that did have valid records been thoroughly searched?

The quest started by trying to locate the 'less than a dozen specimens' known.[10] Since, in the 1990s, the holdings of only a few, mostly American, museums were available, and those were largely in printed computer catalogues, this was slow-going. Frustrated by the difficulty of figuring out where they might be, I wrote and emailed the museum curators of at least 24 likely major collections, only to be told there weren't any *Athene blewitti* in almost all of them.[11] Of course, from the Ripley paper on the Forest Owlet, it was clear that the NHM had four specimens, and I soon arranged to revisit the collection, partly to work on this species.[12] Dr Prys-Jones had kindly arranged to have the single specimen collected by V. Ball in Orissa, which Dr Nigel Collar of BirdLife International had located in Dublin, sent to the NHM for my study, so I was able to examine five specimens there. Upon examining the plumage and taking multiple measurements of each and comparing them with all the forms of Spotted Owlet, I became convinced that the Forest Owlet was indeed a valid species and that none of the few existing identification materials had accurately depicted or summarized the distinctions between it and the Spotted Owlet. To this dataset, I was soon able to add the single American Museum of Natural History, New York, specimen and the Museum of Comparative Zoology, Harvard, specimen of *blewitti*, which further validated this view. This meant that observers could hardly have known how to identify the Forest Owlet up to that time. I commissioned renowned bird artist Larry McQueen to paint a plate comparing the two species, thinking that we could use this in aid of efforts to track the species down. I also thought that, given the lack of accurate information on identification, more misidentified Forest Owlet specimens might come to light in museum collections. Given the huge collecting effort that took place in India, mainly in the nineteenth and twentieth centuries, it seemed unlikely that only seven specimens could ever have been collected (and only over a thirteen-year period between 1872 and 1884) of an Indian peninsular species, especially one that is said to sit out in the sun on the tops of bare trees. However, ultimately no more specimens were ever found to my knowledge. In addition, the more recent sight reports, with accompanying photos, all proved to be mistaken. I had been discussing aspects of this with Dr Collar, and we began collaborating on a paper on the historical record and identification of the Forest Owlet,[13] as well as one on an evaluation of the validity of the data associated with the Meinertzhagen specimen of the Forest Owlet from Mandvi.[14]

The latter investigation into Meinertzhagen's *blewitti* specimen proved more challenging than many others involving Meinertzhagen specimens, because it was clear from the outset that Meinertzhagen's owlet did not match any of the others in style, and because it had obviously been remade some time subsequent to the original preparation. Its remaking was clear from the clean straight edges of the skin along the incision, which would have shrunk back against the stitching in an original preparation; the old disused needle holes around which skin had shrunk; and the sharp-edged neck skin folds, which bore witness to post-drying neck-shortening. And the Meinertzhagen specimen was the cleanest and nicest-looking of the lot, which could have been taken as a testament to the skill of the preparator, but it did not match even remotely other specimens in Meinertzhagen's collection (except some we found later, which had also been remade). So, the Meinertzhagen specimen, about which we initially had suspicions based on what we already knew about his important South Asian bird records, had clearly been remade! That by itself didn't mean it was stolen, but the remaking could well have been an attempt to cover up such a deed. But where could he have stolen it from? Not for lack of trying, we could find no evidence for a missing specimen and seemed to be at an impasse.

It didn't make sense that Meinertzhagen would have collected an extremely rare bird from a place where no one else ever got it, when he had no other specimens at all from there or that time period. His diary didn't mention his having left Bombay to go to Mandvi; in fact it states that he was busy with intelligence work in Bombay during that time. Even if he had gone to Mandvi, it would have taken him considerable time and effort (including on foot or horseback) to get there and back. Why would he have gone to this backwater, collected just one very rare bird, and then never published anything on this triumph? As far as we could determine, the existence of Meinertzhagen's Mandvi *blewitti* specimen was unknown to ornithology until published in the third volume of Ali and Ripley's

Handbook of the Birds of India and Pakistan in 1969, two years after Meinertzhagen's death.[15] That did not stop it from being subsequently mentioned without question in the literature at least sixteen times, though.

Still, there were no leads on where Meinertzhagen could have obtained his *blewitti* specimen. Then one day at the NHM, while searching for a possible match to an oddly prepared, long-necked parrotbill in the Meinertzhagen collection, I was paging through the register of James Davidson's collection. Davidson had collected four of the six non-Meinertzhagen *blewitti* specimens we knew about up to that point (which had come into the NHM as part of the giant Hume Collection, though two were subsequently exchanged to US museums), and his specimens tended to have long necks like that parrotbill. The Davidson register, unlike most, did not have the specimens listed in taxonomic order, so I had to go through it tediously, line by line. I was starting to lose focus when I ran across a listing for *Carine brama blewitti* (*Carine* being a synonym for *Athene*). I turned the page, thinking it was one of the specimens we already knew about. Then it hit me – all the Davidson *blewitti* specimens had come in through the Hume Collection .... Could this really be a long-missing specimen that had been appropriated, remade and relabelled by Meinertzhagen? There was no indication in the register that it had been lost, traded or sold, and it hadn't been present in the NHM collection in the mid-1980s, when all the extinct and exchanged materials were listed.[16] And Ripley and Ali believed that they knew of all the specimens in existence by 1975. We thought that this newly located listing for a missing Davidson specimen must refer to what was now labelled as the Meinertzhagen specimen, but weren't sure how to verify this, since it had been so radically remade.

Even so, clues started coming in, like the much-shortened neck, which would have gone a long way towards transforming a long-necked Davidson specimen into the much better-looking short-necked Meinertzhagen one. Another clue could be seen on X-rays: Davidson, who must not have been well-tutored in bird specimen preparation and seemingly worked up his own procedure, retained the entire humeri (upper wing bones) in his specimens, unlike most other preparators. Meinertzhagen's specimen now has neither the humeri nor the proximal radius and ulna, but the X-rays show that the skin had dried around where these bones once were (and bone removal facilitates remaking of specimens). Another of the several clues tying Meinertzhagen's specimen to the missing Davidson one was that Davidson (alone among the many collectors in Asia whose styles we have studied) often used a method of tying the wings together over the back, but under the feathers. Such a 'wing string' was not immediately obvious in the Meinertzhagen specimen, but a more careful search showed that it had been clipped, though both ends were still there, right where Davidson would have stitched it in.

The nail in the coffin, though, was when we received permission to have J. Philip Angle, then at the National Museum of Natural History, Smithsonian Institution, and a noted specialist in the removal of skeletal material from rare study skins, examine the specimen internally for clues. Since the original stuffing had been removed and the specimen degreased during remaking, we weren't sure what he might find; we were hoping for a few remaining original fibres. But his very first opening produced a wad of greasy cotton from around the wing bones. This was another of Davidson's preparation quirks: the unnecessary wrapping of the wing bones with cotton. Very few other preparators did so, and yet Meinertzhagen's *blewitti* retained a bunch of the greasy cotton in the tight carpal joint area, from which it hadn't been removed or cleaned. Since the National Museum of Natural History (NMNH) is very close to the FBI building and the museum's staff had connections there, we were able to have the cotton from the Meinertzhagen *blewitti* compared by FBI staff member D. Deedrick to a sample from another owl specimen (collected by Davidson within two days of his missing *blewitti*), and they were found to lack significant differences.

Around this time, we also received permission to have some osteological elements removed from one of the NHM specimens by Angle. I had already noted the much broader tarsi of the Forest Owlet compared to those of the Spotted Owlets, but taking this measurement on study skins and X-rays was not nearly as convincing as seeing and measuring the extracted tarsometatarsus, which differs remarkably in its stoutness from the gracile ones of the Spotted Owlet. In addition, the broader head of the Forest Owlet was borne out, as were some unexpected but striking differences in the skull. We considered then that these differences confirmed that Hume's genus *Heteroglaux* should be used instead of *Athene*, but this has not been widely adopted, although a later paper made the case based mainly on osteology.[17]

In trying to piece together the historical record of the Forest Owlet, reviewing every likely publication for clues was necessary, and fortunately the Ripley office had an exhaustive reprint library that made it possible. But the clues were very few! Once I geo-referenced the few known localities (not so easy back then), it became clear that (putting aside the dubious Meinertzhagen locality) the species was only known from four localities, two western ones (Shahada and Taloda) in northern Maharashtra, and two eastern ones in Orissa (Kharial) and eastern Madhya Pradesh (Phuljar). The scant literature about occurrence indicated that one was found in a mango grove along a river, and another source stated that it was found in the 'densest forests'. But subsequent sources indicated and even mapped its broad occurrence in hills, or even stated that it occurred in mountains. However, the actual localities from where the specimens came were in the lowlands, and Davidson, the collector of most of the specimens, stated that he did not think it was found in the higher hills. Thus, I concluded that searches needed to be done in plains forest.

From the literature I could find no evidence that anyone had actively searched for the Forest Owlet in the few verified historical sites since the 1880s, and so I started to formulate a plan to do exactly that. Ben King and David Abbott joined in the quest. Ben, because he was on the trail of Asia's rarest birds (having seen almost all the others) and because he had special expertise in owls, while David had a lot of owling experience, though it was his first trip to India. I was pessimistic about our chances though and thought that at least such an attempt would allow documentation of the status of the habitat and the feasibility of a thorough follow-up search. Colleagues at the Smithsonian's NMNH were pessimistic, too, variously giving us a 1 in 10 to 1 in a 100 chance of success.

In preparation for the search, I tried to obtain remotely sensed data on remaining habitats near the historical sites, which at the time was available only to IT specialists, who could share the data upon payment. Just in the nick of time, a packet arrived from D. Johnson of the Smithsonian, containing two printouts of the Taloda/Shahada area, one with the note 'Best bet is the hills north of the river. There's lots of vegetation, most likely cultivation, between the river and the hills. Nothing much south of the river.' This was our best hint thus far in terms of where to search around the western localities. As for the two eastern localities, all we knew was that one was shot in a mango grove near a river, and the other in the 'densest jungle', and that some of the specimens at least had been collected in broad daylight. We knew nothing about the voice of the Forest Owlet, but I think everyone sort of assumed it would be at least broadly similar to that of the Spotted Owlet, and in fact

Ali and Ripley tried using the Spotted Owlet playback in their unsuccessful search for the Forest Owlet.

With not much to go on we started around the eastern sites on 13 November 1997, looking by day and by night, wherever we could find forest.[18] We encountered a good spectrum of the common birds of these areas, although sometimes we were left with the empty forest feeling, and of course at other times we couldn't find any forest. On at least two days we had to abandon our plans due to unbridged rivers. We showed a colour Xerox of Larry McQueen's painting to a few locals and asked if they knew about the Forest Owlet, but this didn't produce helpful information, although we were told the Forest Owlet had once occurred in the area and was regarded as an evil omen that had to be killed. I initially felt that the eastern sites would be the most likely ones for a rediscovery, but we found nothing of importance except some tangential findings, such as my first realization that Jungle Nightjar and Grey Nightjar (then considered conspecific) differ quite strikingly in song.

After several days of fruitlessly searching in the east, we reluctantly headed to the western sites. I had assumed that the eastern sites would be less impacted by anthropogenic factors and thus be more likely to still have populations of the Forest Owlet, but we had agreed to search all four sites. When we finally arrived in the Shahada area, we first went to the Forest Department and talked about our plans with M. Pokyim, the then deputy conservator. We said that we needed to search for the owl in the lowland forest in the area, but Mr Pokyim told us that there wasn't any lowland forest left, due to a resettlement project of people affected by the Narmada Dam, adding that we could drive up into the hills and see for ourselves that there was no remaining lowland forest. That depressing news seemed like the end of our hopes for finding the Forest Owlet on that trip. Then I started to wonder whether, maybe, the Forest Owlet might also be found in the lower hills? After all, very few species in the region are truly restricted to the lowlands. In any case, we decided that we would head up into the hills and bird anyway.

The following morning, on 25 November 1997, we were up and birding along a road in the low hills long before dawn. I especially remember an obliging predawn Mottled Wood-Owl, but other owls were calling as well, with none of them giving unfamiliar notes. Hopes for finding the Forest Owlet grew even dimmer as the morning wore on and the temperature rose. Local people would walk by, and the occasional bus drove past. There was nothing special or promising about this scruffy woodland habitat that I could see. After videotaping a Shikra, I was opening my water bottle when I heard Ben quietly say, 'Look at that owlet.' I dropped the water bottle and grabbed my binoculars. David was on the owlet too, and then I saw it and said, 'It doesn't have any spotting on the crown and mantle.' Since these features distinguish the Forest and Spotted owlets, we all knew right away what we had found.

How did it feel to succeed in our quest? In place of joy, I was surprised to find myself afraid – that the bird would fly off and we would never be able to prove that we had found the Forest Owlet. But that didn't happen, and with my (now primitive-seeming) video camera I was able to creep up on it and get footage that would bear witness to our rediscovery 113 years since the last definite evidence of *blewitti*, in the form of specimens. My footage is shaky and somewhat grainy, and there's a twig in front of the bird's face, but it undeniably showed a Forest Owlet. Meanwhile, David digiscoped the owlet, obtaining somewhat better results than mine. We were over the moon, and Ben was especially hopeful of getting the first sound recordings of the Forest Owlet. Alas, despite revisiting the site over the next couple of days, Ben never got it on tape, though a possible flight call was heard.

Upon returning to Mumbai, I went straight to the Bombay Natural History Society (BNHS) and let Dr Asad Rahmani, then the Director of BNHS, know that we had found the Forest Owlet after a 113-year hiatus. Dr Humayun Abdulali was also there, and he told me he didn't believe that what we had found was the Forest Owlet, because he didn't think it could be identified visually (he was probably influenced by what Ali and Ripley had written). He further said I should have collected it, and when I told him I didn't want to end up in an Indian jail, he replied that he would have gotten me out of jail. Obviously, I wasn't about to put that to the test, but our video was convincing enough and, shortly after we returned home, the Forest Owlet story made the news, with the *Washington Post* carrying an article entitled 'A Tale of Discovery: The

Owlet and the Fine Feathered Filching' on Sunday, 28 December 1997.

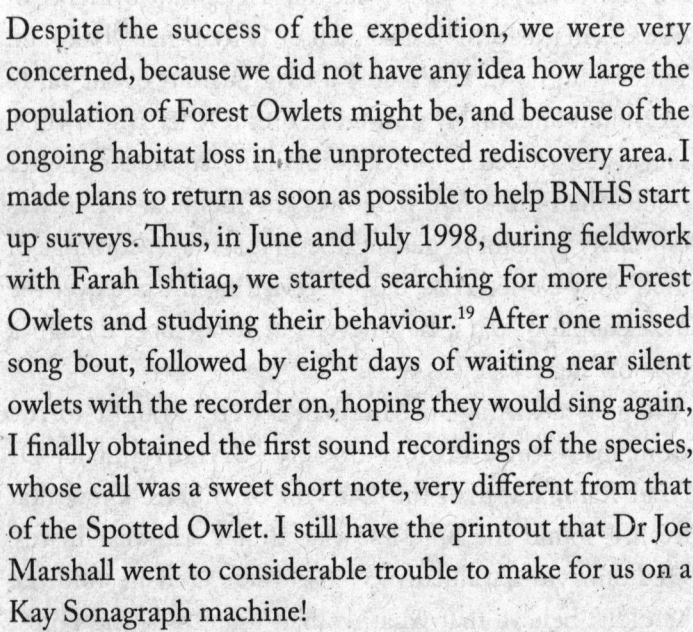

Despite the success of the expedition, we were very concerned, because we did not have any idea how large the population of Forest Owlets might be, and because of the ongoing habitat loss in the unprotected rediscovery area. I made plans to return as soon as possible to help BNHS start up surveys. Thus, in June and July 1998, during fieldwork with Farah Ishtiaq, we started searching for more Forest Owlets and studying their behaviour.[19] After one missed song bout, followed by eight days of waiting near silent owlets with the recorder on, hoping they would sing again, I finally obtained the first sound recordings of the species, whose call was a sweet short note, very different from that of the Spotted Owlet. I still have the printout that Dr Joe Marshall went to considerable trouble to make for us on a Kay Sonagraph machine!

Since then, BNHS has sponsored several projects on the Forest Owlet, and it is a mystery bird no longer. Many excellent photos (vastly better than any of mine!) are online (848 on eBird alone, as of August 2024), as are many sound recordings. Although there are no recent reports from the original rediscovery site on eBird, and it may have disappeared from there, happily the Forest Owlet is now known from multiple protected areas, mainly from Purna Wildlife Sanctuary and Vansda National Park in

Gujarat; from Dadra and Nagar Haveli and Tansa Wildlife Sanctuary and Melghat Tiger Reserve and its vicinity in Maharashtra. To my knowledge, it has not however been relocated in the eastern parts of its former range. Given how distinctive it is in plumage, shape and vocalizations, and given what we now know about its distribution, it is difficult to account for its being overlooked for so long. Though there are plenty of challenges to its survival, its conservation status has been downgraded from Critical to Endangered. Fortunately, unlike the other recently rediscovered Indian species, the Jerdon's Courser, the Forest Owlet is unlikely to disappear again soon.

ABOUT THE AUTHOR

Pamela C. Rasmussen, PhD, has been fascinated by birds since childhood. After completing her PhD dissertation on Patagonian cormorants, she worked at the National Museum of Natural History, Smithsonian Institution, first on fossil birds of North Carolina and then as scientific assistant to the Institution's Secretary Emeritus, S. Dillon Ripley. While preparing *Birds of South Asia: The Ripley Guide* (2005), she led the effort to rediscover the Forest Owlet. Dr Rasmussen has also co-authored the scientific descriptions of eleven new Asian bird species. She taught for over 20 years at Michigan State University, where among other things she co-led 25 study abroad programs. She is a member of the North American Classification and Nomenclature Committee, co-managing editor of the IOC World Bird List, and panel member of the International Ornithologists' Union's (IOU) global taxonomic alignment project, WGAC. She now works as senior research associate at the Cornell Lab of Ornithology, specializing on avian taxonomy.

4

The Chilappan Challenge

PRAVEEN J.

A bird out of time. That could well be the title for this essay. The Banasura Laughingthrush is a bird with the narrowest of ranges, restricted as it is to patches of forest in the high reaches of just a few peaks of the Western Ghats in northern Kerala. Isolated by habitat and location, its seclusion is a double-edged sword; anthropogenic advances are slower, but isolation on montane sky islands puts the bird at risk from the advance of climate change. It takes effort and passion to find such a species, as author Praveen J., who was also one of the first people to suggest that the Banasura Laughingthrush be elevated as a separate species, distinct from the Nilgiri Laughingthrush, demonstrates.

Sports, I am afraid, is not a natural calling for me. Each time I take off on a run, my breath becomes a fickle thing, dissipating far too quickly to keep pace with the spry strides of others. So, while the others with their limitless lung capacity seem to glide through the tracks, I am left to huff and puff, my chest burning from the effort. It's an exhausting reminder that my physical prowess is a far cry from theirs. Yet, there's a defiant streak in me, a stubbornness that compels me to challenge my lungs, to prove to myself that I can compete, that I can be more than just a weary laggard at the back of the pack.

Back in my school days, my adolescent self took on long-distance runs, even though my legs felt like lead and my breath came in bursts. The 800 m was the longest race for 13 year olds, and I entered it with all the trepidation of a soldier walking into battle. I practised, pushing myself until my lungs begged for mercy. I won that race, but the victory felt like a stolen prize, taken from the brink of collapse.

As I grew older, the challenge of the 1,500 m loomed, an unrelenting test of endurance that mocked my limited stamina. I forced myself through the torment of training, grinding away the resistance in my lungs once again until they relented, and I crossed the finish line ahead of the rest. My victory was not just a race won, but also a small victory over the limitations of my own body, a silent rebellion against the inevitability of fatigue.

It was on these long runs, with the exertion of each stride, that I found my new fascination. The mountains became my next hurdle, their peaks promising a challenge that seemed to stretch beyond my physical capacity. The climb was steep, the air muggy, but the summit held an allure that drove me onward. It wasn't just the view or the sense of achievement that enticed me – it was the mountain birds, the chilappans, flitting through the thickets, calling out in a language I could almost understand. Their presence made the climb worth every laboured breath.

And so, the journey to the mountaintops became a metaphor for my struggle against my body's constraints. With each ascent, I sought to prove to myself that I was not bound by my limits, that I could push through the exhaustion and reach the heights I once thought were beyond my reach. The mountain birds were a reward, a gentle nod from nature that my efforts were not in vain. They were a symbol of my victory over fear, over doubt and over the constant ticking of time that seeks to erode our will.

The chilappans, those curious babblers, are a tribe of mid-sized birds that inhabit the towering peaks of the Western Ghats. It is a Malayalam word that means 'babbler' or 'chatterer'. Within the Ghats' rugged range, four distinct species of chilappans have carved out their specific niches. The most renowned among them is the Nilgiri Chilappan, known to frequent the highlands around the hill station of Ooty in Tamil Nadu, where the rolling hills are dotted with tea plantations and forests. Not far behind in repute is the Palani Chilappan, a resident of the lush landscapes of Munnar in Kerala and Kodaikanal in Tamil Nadu. But beyond these familiar names are two lesser-known chilappans, dwelling in the more remote corners of the Ghats. The Ashambu Chilappan inhabits the hills of southern Travancore, a former princely state, now divided between Kerala and Tamil Nadu. Our protagonist, however, is the Banasura Chilappan *(Montecincla jerdoni)*, a bird whose territory is almost exclusively confined to the mountains of Wayanad District in Kerala. This secluded region, with its craggy outcrops, serves as a haven for these enigmatic birds. The Banasura Chilappan, like its kin, is a master of the high places, finding sanctuary in the uncharted wilderness where few dare to tread.

The Banasura Chilappan, also known as the Banasura Laughingthrush, can be elusive, but finding it isn't as daunting as one might think – provided one ventures into its favoured haunts. Its call is distinctive – a fluty melody that carries across the rugged landscape. Though these birds may appear timid at first glance, they are surprisingly bold once found, allowing one to approach with relative ease. In the right habitat, their songs become a constant aural backdrop, with several pairs calling to each other throughout the day, filling the jungles with a lively symphony.

What makes this songbird so remarkable? For a long time, it eluded even the most determined birders, seemingly hiding from the limelight. There's no motorable road that leads to its territory, which makes the journey to find it an adventure in itself. One has to traverse steep paths through dense foliage and over rocky terrain, to reach the sholas, the subtropical montane forests where it resides. The entire habitat is under the forest department, and one needs prior permission to visit and camp. Perhaps there are very few land birds in India with such an inherent challenge of accessibility. It's this elusive nature that has drawn me and so many others to seek it out — a test of endurance and commitment to find this bird.

Restricted to shola forests interspersed with grasslands and rocky outcrops above 1,400 m, the Banasura Chilappan occupies a pristine world. Here, in these high-elevation refuges, it is locally common, announcing its presence with a combination of loud, flute-like songs and sharp chatters. This is a bird that thrives in the seclusion of the mountains, in places where the roads give way to trails and human noise is replaced by the distant calls of other non-human mountain dwellers. Perhaps it is this sense of seclusion and the promise of uncharted territory that continues to captivate those who seek the Banasura Chilappan. The journey to find it is as much about the trek through these breathtaking landscapes as it is about the bird itself. It's a path less travelled, where the reward is not just the sight of a rare bird, but the experience of stepping into a world that feels untouched and is brimming with secrets waiting to be uncovered.

The Banasura Chilappan is a bird that embodies both boldness and subtlety, as well as nature's skill in blending colours and textures into a single, harmonious whole. This well-proportioned babbler strikes the eye with its starkly defined black face, highlighted by bold white eyebrows. The crown of this bird is a slaty brown, while its back and upper parts are painted in soft tones of olive–brown, with a hint of grey. The wings and tail are olive–brown, too, edged with light hues of ochre. The lower throat and breast take on a greyish-white hue, with sporadic streaks of orange–rufous. The rich ochre belly darkens as it nears the tail, and the feathers under the tail present a deeper shade. The black chin and upper throat meld into the black lores and the post-ocular streak, setting off the white supercilium, which extends a little beyond the eye's outer corner. The crown exhibits a subtle, scaly pattern, a mix of darker slaty browns, which give the impression of a cap, contrasting against the dark grey nape and mantle. The back lightens into olive–brown with a hint of grey, with the upper wing and tail keeping to a consistent olive–brown hue. The pale silver–grey feathers around its ear blend seamlessly into the grey of the sides of the neck. The lower throat is a delicate interplay of grey and white, with the breast shaded an ashy grey and streaked from the dark feather shafts. I would like to believe that this chilappan is the most striking among its four cousins, with its bold patterns and contrasting hues setting it apart.

In 2004, I found myself intrigued by the distribution of the chilappans. A detailed perusal of the mountain peaks, ranges and river valleys in the Western Ghats revealed several glaring gaps in our understanding of its distribution. This set me on a journey to explore the so-called 'sky islands' – isolated mountain tops that rise above deep valleys or gaps, which virtually make the specialist flora and fauna of the mountain tops restricted, akin to oceanic islands.

My first target was the Palakkad hills, a range that had loomed over my boyhood expanse, its crest line etched into my memory from countless days spent watching it from the top of my house, and whose foothills gave me my first 300 bird species. Since the early 2000s, in collaboration with Dr P.O. Nameer from the College of Forestry, Thrissur, I have systematically surveyed several forest areas in the Western Ghats of Kerala. Then, in February 2007, we planned a bird survey, during which I managed to trek up to the hilltops of Palakkad hills in two gruelling days and discovered a new population of the Nilgiri Chilappan (variously known as Rufous-breasted Laughingthrush or Nilgiri Laughingthrush).[1] This was one of the toughest treks I have done in the Western Ghats – about six hours of climbing through dense, humid forests. Our pace was halted by a rogue elephant, we lost our way before dusk and had to halt for the night under open skies in a jungle replete with elephants. However, that trip gave me the much-needed satisfaction of having scaled a 2,000 m peak, while also solving one chilappan mystery. It was a hard-fought discovery, but it filled me with a sense of achievement. I felt laureled.

The next puzzle was the Banasura Chilappan. At the time, distribution maps placed this bird from Goa to southern Karnataka in the Kodagu, with no sign of it in Kerala. This lack of focus on Kerala meant the bird hadn't attracted much attention – it was still considered a subspecies, lacking the glamour needed to draw birders and researchers. There was not even an accepted English name – it was sometimes called the Coorg Grey-breasted Laughingthrush or Jerdon's Laughingthrush, but never Banasura Laughingthrush. It wasn't until I learned that Dr Thomas C. Jerdon had first described the species at Banasura peak – a 'high hill separating Wayanad from Malabar' – that I realized the bird's true range.[2] I became convinced that, throughout Wayanad, from Banasura to Vellarimala, the chilappan was likely to be found on the hilltops above 1,600 m.

In December 2007, with C.K. Vishnudas of Hume's Centre for Wildlife Biology and Natural History, Wayanad, I carefully planned a bird survey focusing on the mountain tops in South Wayanad Forest Division. I set my sights on the 2,073-m-high Banasura peak, accompanied by Sandilya Theuerkauf from the Gurukula Botanical Sanctuary. We also dispatched a team led by E. Kunhikrishnan and Satyan Meppayur, two of Kerala's most-seasoned bird-watchers, to explore the Vellarimala highlands.

For those unfamiliar with the legend, Banasura – known simply as Bana – was a demon king of fearsome power, whose many arms were severed by Lord Vishnu, leaving him with just four after an epic battle. This confrontation also saw Lord Shiva on Bana's side, lending his might to the asura king in a conflict that resonated across the heavens. The Banasura peak, to me, resembled the mythical king, with giant arms extending from its massive head. Our four-hour trek to the base of the summit followed one of these colossal arms, starting at the Banasura Sagar reservoir. The journey began in degraded deciduous forests and transitioned into semi-evergreens with patches of grasslands. At 1,400 m we set up camp in a shola patch but there were no signs of the chilappans, though other montane jewels, such as the Nilgiri Sholakili and the Black-and-orange Flycatcher, were detected.

What does this near-mythical landscape of sholas look like? Amidst the undulating plateaus of such Western Ghats sky islands, nestled within the crevices betwixt the hillocks, green enclaves called sholas find respite from the fierce winds of the mountains. In these sheltered hollows, dwarf trees gather in a dense, verdant throng, forming a near-impenetrable tapestry of greenery. The undergrowth brims with saplings and the serpentine embrace of climbers, leaving scarcely any patch of open ground, except for the leaf litter teeming with leeches. The woody trees, though frail and stunted, bear a thick mantle of lichens on their bark, lending them an otherworldly aura. Even in mid-summer, mornings here are frost-laden, and that governs the pace of regeneration. Some botanists deem such shola forests as climax vegetation, the pinnacle of ecological succession. These forests, with their prodigious capacity for water retention, serve as vital reservoirs, their significance magnified for the villages nestled in the valleys below.

The next day, we continued the ascent. As we crossed the 1,600-m mark, the melodious *pe-ko-pe-ko* call of the Banasura Chilappan greeted us. The sight of the bird was exhilarating, and I felt that it looked slightly different from the illustrations I'd seen in books. Throughout our survey, several pairs called from various parts of the mountain, confirming that the prime habitat of the species was still intact and thriving.

Meanwhile, the Vellarimala team also reported a good population of these laughingthrushes, with one member capturing the first-ever photo-set of the species – the only photos for nearly half a decade. Newspapers celebrated our discoveries. I hadn't imagined that we could solve two mysteries within a year. The experience was a testament to the power of exploration and the thrill of uncovering the secrets hidden in the sky islands of the Western Ghats.

Chilappans are not laughingthrushes – at least that's the latest assertion. In 2017, a new genus, *Monticincla*, was proposed for this unique group of montane birds. The name is derived from the Greek word for 'mountain thrush' – a combination of *mons*, meaning mountain, and *kinklos*, denoting thrush. This reclassification underscores the fact that chilappans share a common ancestor with other mountain birds like the Barwings of the Himalayas, suggesting an early radiation from the Himalayan fauna, followed by a complete diversification in the Western Ghats. This makes chilappans a genus endemic to the Western Ghats, separated from the Himalayas by large swathes of inhospitable landscape for those birds. Hence, it certainly must need a different name than the widely used 'laughingthrush'.

However, it is far less well known that the first bird to be ever termed a 'laughingthrush' was the Nilgiri Laughingthrush (or Chilappan), named so by Dr Jerdon for its merry laughing call.[3] So, technically, chilappans were the first laughingthrushes! Now, would that mean all the three score laughingthrushes of the Himalayas and Southeast Asia need an entirely different English name?

In 2010, BirdLife International published a paper describing the 'Tobias Criteria', a new method for designating species.[4] Though that methodology has become largely outdated today, it was a remarkable tool then for rapid assessment of avian species taxonomy. With P.O. Nameer, I used this methodology to propose that both the Banasura and Ashambu Chilappans were independent species. Despite scepticism at the International Conference of Indian Ornithology where we first shared our results, the work was published in the *Journal of the Bombay Natural History Society*, and BirdLife International also accepted the species split.[5] Soon enough, the Sky-Island Project from the National Centre for Biological Sciences (NCBS) confirmed the phylogeny and ancestry of these birds, leading to a four-way species split and the creation of the new genus *Monticincla*.[6] The change was significant because it was the first time that the unique evolutionary history of the chilappans was highlighted.

In fact, the Banasura Chilappan is one of the very few species in peninsular India whose true evolutionary relationships have remained elusive for so long. Despite their discovery in the late nineteenth century by British ornithologists, these birds were repeatedly misclassified and lumped with other Himalayan species. Edward Blyth first described the Banasura Chilappan as *Garrulax jerdoni* based on specimens collected by Dr Jerdon from Banasura peak, suggesting it should form a 'separate division' from the Himalayan laughingthrushes.[7] Yet, ornithologists continued to assign chilappans to various Himalayan genera, such as *Garrulax*, *Trochalopteron* and *Strophocincla*. Authorities of avian taxonomy kept shifting them from one genus to the other and sometimes back. This led to considerable nomenclature instability and reduced attention to the unique characteristics of the chilappans.

It became even more problematic when ornithologists started lumping these species based on their shared morphology. In the early twentieth century, when the subspecies concept was in vogue, the Banasura Chilappan, which shares the grey breast with both Palani and Ashambu Chilappans, was lumped together with these two under one species as the Grey-breasted Laughingthrush! Hence, for nearly a century, the Banasura Chilappan lost its uniqueness as it was considered just a colour variant of the more widespread Palani Chilappan. The lack of ornithological attention for this species was almost entirely due to this misclassification. Subspecies never get the limelight, even now!

Pamela Rasmussen, in 2005, reverted to Blyth's original assignment, moving Banasura Chilappan back to being at least a subspecies of the Nilgiri Chilappan, and she named the species 'Black-chinned Laughingthrush'.[8] This reshuffle, though initially scoffed at, prompted further research into the group's taxonomy, eventually leading to the recognition of the new genus *Monticincla*.

Subsequent morphological analysis revealed consistent differences between the Banasura and Nilgiri Chilappans, particularly in the colour of ear coverts, breast, neck sides and the intensity of the rufous colour in its lower breast. Genetic studies confirmed the reciprocal monophyly of the Banasura Chilappan with the other three chilappans, indicating that all samples of Banasura formed an independent branch in a species tree.[9] Additionally, all four chilappans have allopatric ranges, separated by gaps that have persisted for millions of years, likely due to their weak flying abilities which limited dispersion across these gaps.

Despite these advancements, much remains unknown about the Banasura Chilappan. I was fortunate to update the Birds of the World (a website that collates ornithological knowledge) species account for this bird, and some of my observations remain the only evidence of its vocalization and breeding behaviours. It is believed to feed on insects and fruits, especially wild raspberry and hill guava in the sholas, similar to the Nilgiri Chilappan, but actual data is limited. There was just one song recording until 2012 until I started collecting more, which emphasizes the need for further study to fully understand this unique species.

The taxonomy of the chilappans piqued my interest in their vocalizations. I embarked on journeys to the sky islands inhabited by the Banasura and Ashambu Chilappans, with a simple goal: to capture their songs and calls and compare them with existing recordings of the Nilgiri and Palani Chilappans. Accompanied by Vishnudas, I trekked through the rugged terrain of Vellarimala, set up camp and spent time listening to the lively chorus of the Banasura Chilappans. The birds had already established their territories, with at least six pairs near our camp preparing for the breeding season. This visit allowed us to record a treasure trove of vocalizations, revealing the remarkable variety and plasticity in their songs. These recordings are now safely stored in the Macaulay Library and have found their way into the *Birds of the World* account.

Chilappans are known for their versatility in song, and the Banasura Chilappan is no exception. Initially, its voice was described as subdued compared to the Nilgiri Chilappan, but my field recordings challenged this notion. The Banasura's repertoire is both wide-ranging and intricate, with duets forming the backbone of many of its songs. These duets are usually variable but tend to follow a well-defined frequency band. The true breadth of their variations – whether they indicate individual differences or subpopulation dialects – remains an intriguing question and is yet to be researched.

Perhaps its characteristic song may be described as a variable series of three to six nasal whistles, usually joined by a second bird in duet, giving a hoarse, treepie-like laughing – *urg-urg-urg*. Some of the songs I recorded in Vellarimala were not the common vocalization that others have now recorded elsewhere in its range. Some longer song types include a four-noted, strongly nasal, rather complaining *wo-ko-ke-keh*, followed by a duet chatter, which may continue as a long, rising and less-hoarse crescendo of short, agitated, nasal *aingk* notes. All my recordings were songs, and I did not record any calls. However, its call is rather similar to the Nilgiri Chilappan, which produces a variety of harsh chatters similar to a Rufous Treepie or a Jungle Babbler.

Researchers from NCBS analysed sound recordings of Banasura and Nilgiri chilappans using five key song parameters. They found the Banasura species to have a higher song complexity with longer phrases, yet with considerable overlap in the study parameters. No single characteristic distinctly set their calls apart from the Nilgiri Chilappan, suggesting that a much larger sample size might be needed to draw definitive conclusions. The propensity of these birds to drop phrases and notes also makes this job rather difficult.

In December at Banasura and in January at Vellarimala, the full repertoire of the chilappans was on display. They are generally vocal throughout the day, with a slight uptick in activity during mornings and evenings. Songs and contact calls are delivered while foraging or when alarmed. No songs or calls were recorded while in flight. My observations suggest that each individual possesses multiple song variations, though I found no clear distinction between male and female vocalizations. However, their potential for duets strongly suggests male–female differences in vocalizations.

How do you find one? The highlands of Wayanad – particularly the sholas where the Banasura Chilappans dwell – have historically been off-limits for trekkers. However, tourist resorts are slowly climbing their way up the mountainsides. In 2023, birders discovered a new tourist resort near the chilappan's habitat in Vellarimala, that created an easier access route to this elusive species. Since then, the site has attracted several birder groups, allowing them to trek to these remote spots and photograph this once-elusive bird. It's likely that this site will remain one of the few accessible gateways to the chilappan's habitat for the foreseeable future.

British ornithologists and bird collectors have long been drawn to these sky islands, obtaining chilappan specimens from the mountains of the Brahmagiris and Banasura.[10] Recent molecular studies relied on live specimens from Ambalappara (in the Brahmagiris range) and Banasura. The twenty-first–century discovery of the chilappan in Vellarimala, also known as Camel's Hump Mountains, has been supported by photographic evidence, but the lack of expeditions during most of the twentieth century makes it difficult to assess changes in distribution over time. Since the late 1990s, there has been a concerted effort to map the distribution of all sky-island specialists. The Nilgiri Sholakili, which co-occurs with the Banasura Chilappan in most sky islands, serves as a useful proxy for identifying suitable habitat and elevation. Typically, the chilappan's preferred habitat is highland broadleaf evergreen forests, particularly shola forests with dense undergrowth and their edges. Although it can be found at altitudes as low as 1,400 m, the bird is most commonly seen above 1,600 m, reaching the highest peaks in these sky islands.

What are the chances of finding this bird in Karnataka? The strict altitude preference of the Banasura Chilappan limits the areas we need to search in. Notably, there are no peaks above 2,000 m in Karnataka, and Banasura peak is the northernmost 2,000+ m peak in peninsular India (south of the Himalayas). In search of satellite populations, I trekked through the sholas of various mountaintops in the central Western Ghats of Karnataka, including Pushpagiri Wildlife Sanctuary, Kudremukh National Park and Bababudan hills. Despite historical distribution maps suggesting a broader range, I found no signs of the Banasura Chilappan in these areas. Sightings from the 1990s and 2000s raised initial interest, but subsequent visits by birdwatchers and researchers yielded no authentic records. The absence of supporting documentation for these older sight records makes it difficult to determine whether they indicate a local extinction or were potential misidentifications. The Nilgiri Sholakili, its closest proxy, continues to be reported from these areas, further suggesting that the search effort is reasonable, but the Banasura Chilappan might not occur north of Aralam Wildlife Sanctuary (Kerala). The nineteenth-century specimens collected by William Davison from the 'Brahmagiris' likely came from the extension of the Brahmagiri sky island into Aralam in Kerala, hinting that this bird might not even occur in Karnataka proper. Reports from Castle Rock in northern

Karnataka and Dudhsagar in Goa, much farther north of its known range, have been retracted, strengthening the case that this species might now be entirely restricted to Kerala.

Could it be found in other highlands in Kerala? The Vellarimala block is the largest sky island north of the Nilgiris; the Chaliyar River separates the Nilgiri massif from the high peaks of Wayanad, forming a speciation barrier for chilappans. The pinnacle of Vellarimala forms the boundary between Wayanad, Kozhikode and Malappuram districts, allowing the Banasura Chilappan to be included in the district checklists of all three. The famous Thamarassery Churam (the local word for 'mountain pass') lies north of Vellarimala. Between Thamarassery Churam and Banasura, a couple of high peaks, specifically Kurichiyarmala (1,570 m), were previously reported to house the chilappan. However, recent surveys have found no evidence to support this claim. Other peaks are too low for this species or do not have shola forests. The Periya Churam separates Banasura and Aralam, with no other highland in between. Overall, with the bird's habitat confined to a few specific high-altitude regions, the chances of finding it in any other sky island in Kerala are slim.

So what does all this portend for the bird's future? The Banasura Chilappan has been classified as 'Endangered' by BirdLife International – the only Western Ghats sky-island species with such a high threat status.[11] As its range is perilously small, even the slightest disruption can have devastating consequences. With an estimated population of 500–2,500 mature individuals, each subpopulation holds a critical piece of this bird's fragile existence. Vellarimala offers the largest habitat, about 38 sq. km, while Banasura has a mere 5 sq. km, and Ambalappara, within the Aralam Wildlife Sanctuary, encompasses about 15 sq. km Even in Ambalappara, the population density is lower than the other two, indicating that the actual habitat may be even more limited.

The Banasura Chilappan is most commonly found at elevations between 1,800 and 2,000 m above sea level, exhibiting a density five times higher than that at lower altitudes of 1,400–1,600 m. This indicates a clear preference for montane shola forests at higher altitudes. While no extensive population-level studies have been conducted, surveys in the Banasura area recorded the highest encounter rates. Given its near-endemic status to the Wayanad District, the Banasura Chilappan stands as an iconic species, capable of fostering meaningful connections with the general public, thereby catalyzing robust conservation efforts.

Despite the habitat protections offered by reserve forests where most of its range lies, the encroachment of ecotourism, particularly trekking, poses a growing threat. Imagine the chilappan, perched quietly amidst the dense foliage of the sholas, when its sanctuary is suddenly disturbed by the clamour of human voices and the tread of boots. The main trekking routes at Chembra and Ambalappara lie just a kilometre from the chilappan's remaining habitat, providing a small buffer from the constant influx of visitors. In places like Pushpagiri Wildlife Sanctuary in Kodagu, where trekking routes once cut through the sholas, the chilappan vanished by the twenty-first century – a stark reminder of how fragile this species is.

This was brought home sharply while trekking to the summit of Kumaraparvath, inside Pushpagiri Wildlife Sanctuary. Our team had stayed at Bhattara Mane, the only mid-way basecamp on the climb. As we ascended towards the summit, my initial excitement turned to dismay. Along the way, I encountered several independent groups – a few dozen youngsters – who had camped overnight in all the high-altitude sholas. The sight was disheartening: they had rampantly cut down vegetation for their supplies and lit campfires, leaving behind a trail of destruction. The air, usually filled with the dawn chorus of birds, was instead dominated by the ruckus of human activity, echoing throughout the sky islands.

Luckily, none of the sky islands where the Banasura Chilappan is found have extensive commercial plantations, and hence the threat from environment pollutants, like pesticides, is likely to be less. However, this is only a surmise, and would need robust studies to confirm the status quo. Overall, the populations of the Banasura Chilappan are suspected to be in a slow decline owing to the loss and degradation of habitat; however, quantitative data is lacking.

Do we know enough to conserve this species? True, the habitat and altitude preferences of this bird are well-known, and that straightaway limits the focus to a tiny sliver of the Western Ghats. However, we know practically nothing about its life history – no details about its breeding, survivorship or demography. Its interaction with its habitat, at best, is mere conjecture. Its lifespan is assumed to be five years – based on data from other chilappans. There is hardly any information on its non-breeding habits. Hence, what management practices are needed to ensure the continuity of a viable population are based on good intentions and not hard science. In addition, fire ravages all types of forests in Wayanad. What is the response of this species to wildfires? It is well-known that small populations are threatened by persistent wildfires but several other species rebound with effective fire management. Shola forests of the Western Ghats have been invaded by invasive species like *Lantana camara* and *Ageratina adenophora*, with the former having a demonstrable impact on tree regeneration. There could be linkages of forest fires to these invasive spreads, and understanding this connection should be a research priority.

The unique set of issues surrounding the species pose a challenge but also an opportunity. The range of the Banasura Chilappan extends across an area that is just shy of 100 sq. km, offering a rare opportunity for focused conservation. Most of its range has not been encroached upon, and hence it is comparatively easier to prevent further encroachment. However, barring Aralam, neither Banasura nor Vellarimala fall inside protected areas. In fact, they are classified as 'vested forests', providing some legal challenges to include it under a protected area.[12] The lower western slopes of

Banasura peak fall inside the Malabar Wildlife Sanctuary but the higher elevations were excluded during the sanctuary's notification. There are no sanctuaries adjacent to Vellarimala. A proposed 6.8 km tunnel connecting Kozhikode and Wayanad, which will drill through the mountains of Vellarimala, has been a major cause of concern. Landslides, which could potentially disrupt the sky islands, are predicted, due to the deep excavation required for a tunnel. The presence of this bird was the prime reason behind recommending Camel's Hump Mountains as an Important Bird and Biodiversity Area (IBA). Though this does not guarantee any legal protection, it does bring in international attention.

As climate change pushes highly mobile species such as birds, to higher elevations, one wonders whether the chilappan will go extinct if its mountaintop sanctuary becomes inhospitable. These changes are grim reminders that even seemingly robust ecosystems can deteriorate, and the Banasura Chilappan's future hangs in the balance. The fate of this bird, so bound to the high peaks, is a poignant reflection of the mounting pressures on our natural world. While the odds seem stacked against the species, there is still hope. Conservation efforts, however challenging, may yet stave off the inevitable, giving this small bird a fighting chance to endure in the shadows of the mountains it calls home.

ABOUT THE AUTHOR

Praveen J. is a computer engineer by training and a passionate birder. Hailing from the Palakkad Gap in the Western Ghats of Kerala, where he spent much of his boyhood birding, he developed a deep love for studying birds and their distributions. Praveen harnesses the power of 'bird networks' and is deeply involved in various citizen science initiatives, particularly with eBird, to monitor bird populations across India. He serves as the chief editor of the *Indian BIRDS* journal, which publishes research on bird distribution and identification. Recently, he hosted a comprehensive update to Simon Dillon Ripley's *Synopsis of the Birds of India and Pakistan* as a website (www.birdtaxonomy.in). Currently, Praveen works as a scientist at the Nature Conservation Foundation in Bengaluru, where he leads their bird monitoring team.

5

Following in the Footsteps of the Elusive Masked Finfoot

SAYAM U. CHOWDHURY

Like the Jerdon's Courser, the Masked Finfoot is another species we have failed – the enigmatic wader has now disappeared from across most of its range. It is likely going down fighting the increasing levels of salinity linked to climate change across its habitat arc in the Bangladesh Sundarbans – it is yet to be found in the Indian Sundarbans, though there are four historical records from Northeast India. Geopolitics too plays a role in the survival of the species – little is known about what is happening within Myanmar, a major location for the species, and foreign birders have not been able to venture into the country since 2009. In its last breeding stronghold of the Bangladesh Sundarbans, the population of the Masked Finfoot may be down to less than 100 individuals. Sayam U. Chowdhury's quest to find the species was a test of patience and persistence, gently pursued – as his essay illustrates.

I grew up immersed in the haunting tales of man-eating tigers penned by the great Jim Corbett and the captivating adventures of Sálim Ali as he documented little-known birds across the Indian subcontinent. Their vivid depictions of the unique, untouched, unbroken and unfathomable wilds of India and – Bengal, in particular – left me fascinated to my core.

As a teenager, I fantasized about wandering through our rainforests, conquering our dwarf mountains, sailing across the cryptic, narrow creeks of the Sundarbans, witnessing the splendour of the sunset by a winding river and listening to the enchanting calls of night owls in our sal forests.

The solemnity of being close to nature, the remarkable colours of birds and butterflies and the liquid beauty of the rolling rivers of Bengal have been genuinely praised by Rabindranath Tagore, Jibanananda Das and many other Bengali polymaths in their poems, songs and novels. Cherishing those beautiful verses and stories of endless wilderness in my mind and soul, I ventured out to explore and transform what I had read in those dusty old pages into reality. I envisioned a dark forest where sunlight would be blocked by the top canopy, where moisture would be trapped in the lower levels, while the soggy forest floor fostered the flourishing of insignificant life and where birds would fly without fear. I believed that some of our forests, such as the Chattogram Hill Tracts (CHT) in southeast Bangladesh, would be so thick that one would be able to see the sunrise or sunset only from a few vantage points. The foliage would be so dense and the trunks of trees would be so huge that they would obscure the horizon. Treefall or fire would poke a few holes in the forest, but mostly the tree canopy would be closed for hundreds of square miles. Tigers, elephants and rhinoceros would rule the ground, and birds of prey would master the sky.

The reality unveiled something different – most forests across Bangladesh have been thinned, trimmed, tortured and, thus, threatened; sunlight slickly touches the ground, and the foliage on the forest floor is so dry that it crunches and crumbles under your feet as you walk through it. With the destruction of forests and the pressures of hunting, we lost tigers, bears and gaur in the mid-nineteenth century. Reports of the Sumatran Rhinoceros in northeast Bangladesh date even further back to the eighteenth century – and they have not been seen since – while the last record of the Great Hornbill from Lawachara National Park (Moulvibazar, Bangladesh) was in 1990.

Rural Bangladesh is still green with miles of paddy fields, but not the kind of green I wanted to see. There are hardly any trees left, and those that remain are mostly non-native acacias and eucalyptuses planted in unflattering rows. It is hard to believe that the dry plains of northwest Bangladesh were a haven for megafauna a few decades ago; there were Striped Hyenas and Blackbucks until the end of the nineteenth century, and Gray Wolves and Nilgais until the 1940s.

Similarly, all those stories from the past fascinated me until I travelled through Bangladesh's rivers and streams, freshwater lakes and marshes, haors,[1] baors,[2] and beels,[3] and estuarine systems with extensive mangrove swamps. To my disappointment, I found that most of the inland wetlands had been converted to agricultural fields and other development projects, pushing once common species to the last remaining pockets of wetlands in northeast Bangladesh. The Cotton Pygmy Goose that once occurred in every village and wetland has now become a rarity. I imagine that the Pink-headed Duck was also common in the eighteenth century, but is now globally extinct, with the last confirmed record near Beanpole, Bangladesh, in 1923. The incredible Indian Peafowl occurred in our sal forests and existed there until the early 1980s.

But all is not lost from our land. The forests of Bangladesh still echo with the magical calls of the endangered Hollock Gibbons, and Elongated Tortoises still rustle the leaf litter to find fallen fruit. The mighty Sundarbans support our last tigers, and its murky waterways still hold the critically endangered Masked Finfoots. Elephants continue to silently wander through the *jhum*[4] lands, and some claim that tigers still roam the remote parts of the CHT. Asiatic Wild Dogs or Dholes hunt in packs, Clouded Leopards skulk and the Great Hornbills still make a whooshing sound while taking off high in the canopies of the Kassalong and Sangu-Matamuhuri Reserve Forests within the CHT.

Against this background, I wanted to understand the current state of some of these threatened animals, the problems they were facing and how we could solve those using science. As part of my undergraduate thesis, I began looking for the Critically Endangered Spoon-billed

Sandpiper. At the time (in 2010), there were only a few records of the species, but my colleagues and I managed to find a globally significant population wintering in Bangladesh. This population was then facing immediate threats – such as hunting – and we worked towards addressing these. But this is not a story about the Spoon-billed Sandpipers but a bird that was completely ignored by the global conservation community, a bird about which we knew very little, much less than the Spoonies.

> *There stands a forest*
> *Where rivers flow in rivulets*
> *Roots feed on tides and grow tall with joy*
> *Ferns build nests on trees and befriend the lichens*
> *Underneath the canopy, crabs bury fallen leaves in secret*
> *Mangrove Pittas hop around and compose happy melodies*
> *In between all these, the tiger and the deer roam*

Imagine a moonlit night, a boat cruising gently through a labyrinth of tidal creeks with walls of mangroves on both sides, while the eerie calls of fish-owls echo through the dense forest. In this enchanted realm, amidst the shadows of tigers and the stealthy movements of crocodiles, lies the Sundarbans – a place where mysteries are woven into the very fabric of the mangroves. In the heart of this magical sphere, the saga of a lesser-known species unfolds – a feathered phantom battling for survival. In the vastness of the Sundarbans, Keora trees stand like sentinels, their branches supporting a green cloak of foliage which shields the forest against all odds. It's a world where time seems to stand still, where the only movement is the gentle sway of the boat over murky waters. But within this apparent stillness, a storm has been brewing – a storm of extinction for the Masked Finfoot (*Heliopais personatus)* and I only comprehended the depth of this possible storm when I first began to look for the finfoots in the Bangladesh Sundarbans in 2011.

Finfoots are waterbirds that inhabit the increasingly fragile borderlands of our waterbodies and wetlands. They are rather unique to look at, resembling grebes or ducks as they move through water, though their graceful necks give them the jizz of cormorants and their brightly coloured beaks harken of hornbills! Three different species of finfoots are found across the world – each with its own distinct habitat and distribution. The African Finfoot resides in the streams of tropical Africa, within woodland areas. The Masked or Asian Finfoot is scattered across South and Southeast Asia. The Sungrebe or American Finfoot inhabits the tropical regions of Central and South America. These birds are part of the *Heliornithidae* family – a small group of tropical birds distinguished by the webbed lobes on their feet, resembling those of grebes and coots. Recent studies suggest that the *Heliornithidae* family is closely related to the *Sarothruridae* (flufftail) family, forming a clade[5] that is sister to the *Rallidae* family, which includes rails and coots. Recent morphological and genetic studies point to a close relationship between the finfoots (and other members of the *Heliornithidae* family) and birds of the Gruiformes order, which includes cranes, rails, crakes and relatives.

At that time, in 2011, the global population of the Masked Finfoot was believed to be fewer than 1,000 individuals, with Myanmar holding a majority of this number. The intricacies of its existence were severely threatened by habitat destruction, river disturbance and the gloomy shadow of hunting. In a world preoccupied with the grandeur of Bengal tigers, elephants and rhinos, the Masked Finfoot struggled to find a place in the limelight – no one cared.

Undeterred, we embarked on a two-month expedition, navigating through the labyrinthine creeks of the Sundarbans. The days were long, starting at dawn and ending after dusk. Armed with determination, we searched for the elusive nests of the Masked Finfoot. Our journey took us to the far reaches of the Sundarbans, from Chadpai to the Sarankhola range, covering over 100 sq. km. The Sundarbans, the world's largest mangrove forest, sprawls across the coastal regions of south-west Bangladesh and West Bengal, India. This vast expanse, covering approximately 6,017 sq. km in Bangladesh alone, holds immense ecological significance and is recognized as a Ramsar[6] and UNESCO World Heritage Site. Representing the largest protected area and reserve forest in Bangladesh, the Sundarbans are a critical hub of biodiversity and local economy, encompassing diverse ecosystems from dense mangrove thickets to intricate waterways. In addition to its rich biodiversity, the Bangladesh Sundarbans hold significant socio-economic importance. It's a lifeline for over 3.5 million people who depend on it for their livelihoods. The Bangladesh Forest Department's records show that in the fiscal year 2011–2012, 4,800 tons of fish, 165 tons of honey, 1,600 tons of firewood and 82,700 tons of Nipa Palm were harvested, and 1,83,600 tourists visited the area.

Looking for the finfoot required a relentless pursuit in this mighty forest – a quest that demanded not just physical endurance but the resilience to face the uncertainties of nature. There were days of hope when we caught glimpses of finfoots foraging during low tide. And then, there were days of despair, when nothing was found except for the footprints of spotted deer and the calls of Mangrove Pittas. Sometimes it rained all day and night, or the weather remained unfavourable to conduct an unbiased survey. The terrain presented its own challenges, too, and required us to leave our larger boat and climb into a small dinghy and silently search the never-ending *khals* or narrow creeks along the eastern side of the Sundarbans. Our dinghy traversed up and down the canals, at times with nothing but hope in our hearts. Through the heat of July, the monsoon rains in August, the starting of Ramadan and breaking of the fast, our search for the bird continued. Yet, in the face of adversity, we pressed on, fuelled by the desire to unravel the mysteries of this elusive waterbird.

One of the major challenges for us was to understand where exactly in the riparian vegetation we should look to find the nests, followed by the issue of covering a large area within a short period of time, with utterly limited resources. The first issue was significant – despite reading all the available literature and inspecting photos of nests previously found in the Sundarbans, we found it hard to grasp the entire picture. It was only overcome when we found the first nest, after which our understanding improved significantly. Observing the nest in its natural setting provided invaluable insights, particularly regarding its positioning relative to the water level at high or low tide. We found all nests on the outer edge of vegetation located roughly 1–2 m above the water level at high tide.

The journey was marked by both triumphs and tribulations. We discovered thirteen nests, two active and eleven from the previous year. To understand the bird's breeding biology, we set up a camera trap near one of the active nests; for the very first time ever, the finfoot's nocturnal behaviour at its nest was observed and documented. For intensive observations, we also set up a hide on a wooden boat moored 20 m away from the nest and took turns monitoring the nest for twelve hours each day, from 6.00 a.m. to 6.00 p.m., and studying their behaviours and feeding habits. Some of us developed a unique bond with this pair, particularly with the female, who spent the majority of her time either incubating or foraging around the nest. One day, a Shikra attempted a daring raid, threatening a hatchling that had ventured too close to the water. The fierce mother, however, emerged as the guardian, thwarting the predator's advances. Witnessing this behaviour instilled in us a deep sense of respect and admiration for the female of this particular nest. Not all chicks were so lucky; in 2014, another nest experienced predation by a Changeable Hawk Eagle, emphasizing the vulnerability of the nests in the Sundarbans ecosystem.

Over three remarkable years of study – 2011, 2013 and 2014 – we discovered five active nests, each holding its own unique tale. Clutch sizes varied from one to six. At one nest, two eggs hatched successfully, while all eggs at another nest were predated. The fate of eggs in the remaining three nests remained undetermined. The nest where two out of three eggs hatched showed an interesting variation in intra-clutch egg sizes, where the smallest egg failed to hatch.

In one of the active nests, which was found on 11 July 2011, the chicks hatched six weeks later, on 23 August. Imagine the enchantment of encountering delicate hatchlings, just a day into the world. These tiny wonders, adorned in dark grey above and light grey below, added life to the creek scenery. A significant moment unfolded on 30 July 2011, when we saw two chicks (of a different clutch) with an adult female foraging in light rain, before vanishing into the grey and green horizon of the creek.

From the nineteen-day observation carried out at the nest, a minimum incubation period of over three weeks was evident. Initially, both the male and female of the species shared incubation duties. Incubation changeovers were accompanied by bubbling calls, and the female consistently incubated throughout the night, occasionally changing positions. During this time, the Masked Finfoot pair also explored the creek, venturing up to around 500 m away from the nest. Notably, the incubation pattern shifted in the final nine days; the male departed before the chicks hatched and the female took charge; once he ceased incubation, the male began foraging approximately 700 m away. Foraging trips focused on gathering small crabs (81 per cent) and shrimps (19 per cent) during low tide on muddy banks.[7]

The discovery of active nests during 2013–2014 revealed a consistent breeding pattern, spanning from mid-June to mid-September. This finding aligns with previous research on the span of the breeding season's duration. All breeding females exhibited an unusual yellow fleshy knob above the upper mandible, not previously documented in female Masked Finfoots. The fleshy nobs observed on certain birds during the breeding season are often ornamental structures used in courtship displays. Typically, these prominent and colourful caruncles or wattles are found in male birds, and serve to attract mates or establish dominance within their social hierarchy. However, the reason for the presence of such ornamental features in female finfoots is unclear.

The story of the finfoot that emerged from the study was one that highlighted the harsh realities of the bird's fragile existence. Our observations provided a snapshot of the risks and survival strategies used by Masked Finfoots during their critical breeding period in the Sundarbans.

Even as we immersed ourselves in the study of the finfoots, along the way we encountered the human dimension of the Sundarbans – the fishermen whose lives are intricately linked with the ebb and flow of the mangrove's secrets. Interviews with these fishermen revealed a complex mosaic of relationships; many – 56% of the 68 fishermen interviewed – had hunted or tasted Masked Finfoots, unaware of the consequences of their actions. Many fishermen recounted capturing Finfoots themselves or stumbling upon their nests while setting up their *charpata jaal* or fishing nets along narrow streams in the Sundarbans. Typically, *charpata* fishermen install long fishing nets along the banks of khals during low tide and retrieve the catch after high tide. While affixing such charpata nets near nests along the creeks, some fishermen inadvertently flushed out incubating finfoots, and returned at night to seize the unfortunate bird, its eggs, or chicks from the nest. To me, the Masked Finfoot has become not just a biological enigma, but yet another example of the delicate balance between conservation and livelihood. Local fishermen do not depend on finfoot meat or eggs, but it is a delicacy, an added source of protein for an incredibly poor community.

Between 2011 and 2014, we completed our initial research on the Masked Finfoot in Bangladesh. In 2020, our findings formed the basis for a global population estimate, and the impact of this work was underscored in 2022, when the species was uplisted from Endangered to Critically Endangered, through a comprehensive global review done by me and my colleagues. More than twenty years ago, the Masked Finfoot was a common sight in Thailand and Malaysia during its non-breeding season. However, the swift deterioration of its breeding grounds in Myanmar and Cambodia – primarily caused by habitat destruction, hunting, pollution and disturbance – has led to a situation where there may now be fewer than 300 individuals of this species remaining globally. In 2023, a paper that I presented on our study at the 11th Meeting of Partners of East Asian-Australasian Flyway Partnership led to the establishment of the Masked Finfoot Task Force, through which we hope to make a difference to this species, not just in Bangladesh but also in other countries in South-east Asia that constitute its global range.[8]

The survival of this species in other parts of South-east Asia is completely linked to the availability of pristine forested wetlands. However, it faces increasing threats to its habitat in Myanmar, Laos and Cambodia. Human activities, such as agriculture, fishing and hunting, are rapidly transforming these critical areas. Oxbow lakes in Myanmar, once untouched, are now subject to extensive fisheries programmes. The construction of dams, particularly on the tributaries of the Mekong, further reduces the availability of low-lying riverine wetlands, impacting core finfoot habitat. Preserving the remaining populations of this elusive wader necessitates concerted efforts, both politically and within local communities, and entails the conservation of existing low-lying forested wetlands, especially in Cambodia and northern Myanmar.

The two known breeding populations of the Masked Finfoot in Cambodia exist in protected areas, namely, Kulen Promtep and Chhep Wildlife sanctuaries. However, even these sanctuaries are not immune to (significant) threats like hunting, egg and chick collection, and the clearance of forested riverine vegetation – a problem particularly prevalent in Cambodia. Additional challenges include the accidental entanglement of birds in widely available monofilament fishing nets and the use of poisons while fishing, a practice also followed in the Bangladesh Sundarbans. Managers of these protected sites often grapple with limited capacity and resources to enforce regulations on natural resource harvest. One upside of human activity is the fact that Masked Finfoot nests are frequently discovered by local communities during routine activities such as fishing.

To tackle these challenges, suggested protective measures include restricted access during the breeding seasons and temporary limitations on gill net use. Furthermore, large-scale education and outreach campaigns aimed at local communities are crucial for emphasizing the global significance of the Masked Finfoot and the preservation of forested wetland habitats. Despite the enduring relevance of detailed conservation and research recommendations from BirdLife International, their infrequent implementation over the last two decades means there is greater urgency – now more than ever – to take action for the conservation of the bird and its habitat. This involves developing national action plans for key countries, such as Bangladesh, Myanmar, Laos and Cambodia, to prevent further population decline and the possible extinction of the Masked Finfoot.

Looking back over more than a dozen years devoted to the cause of the enigmatic Masked Finfoot, the memories are not of hardships but of the positive moments – the discovery of nests, the attachment to the finfoot families and the steps taken to improve their condition. It's a journey that goes beyond the confines of scientific exploration; it's a narrative of connection, of humans and nature intertwined in a delicate dance. The Sundarbans story reflects our complex connection with the natural world, capturing the ongoing interplay between humans and the environment. It's a tale that spans borders and eras, and portrays a nuanced dialogue of conservation that echoes within the mangroves, expressing the shared concerns of the finfoot and the diverse life inhabiting this mystical region.

And so, our journey through the Sundarbans comes to an end, leaving us with a profound sense of awe and responsibility. This is not just the story of the Masked Finfoot; it's a narrative of humanity's role in the delicate tapestry of nature. As we navigate the intricate khals that lattice the heart of the mangroves, we carry with us the echoes of a call – a call to protect, to preserve and to ensure that the feathered phantom of the Sundarbans continues to dance in the murky waters of its home.

ABOUT THE AUTHOR

Sayam U. Chowdhury has been working on the ecology and conservation of endangered species in Bangladesh and across South-east Asia for the past fifteen years. He holds a master's degree in Conservation Leadership from Cambridge University and is currently completing a PhD in Conservation Science at the same institution. Sayam works as the coordinator of the Spoon-billed Sandpiper Task Force and collaborates with BirdLife International partners across the region. Sayam is particularly interested in developing science-based solutions to conservation problems faced by threatened species in Asia. In his free time, he enjoys photography, nature sound recording, poetry and reading.

6

Owl of the Emerald Island

SHASHANK DALVI

The Nicobar Scops Owl presents a conundrum for seekers of rarity – it is an owl found in just two islands in the whole world, but once you get there, it is extremely common. What makes the bird special is that, for the longest time, it was an overlooked species, understudied due to lack of specimens. The way Shashank found it – making the connection to related species (no spoilers here!) – tells us how you need to have the right piece in place to unlock the puzzle of a 'lost' bird, and that piece may be biogeography. Even with this puzzle solved now, the Nicobar Scops Owl remains a difficult bird, given the logistics involved in getting to its habitat. But once you get there, it may well be one of the fastest lifers you tick off!

On 15 October 1963, Humayun Abdulali boarded a flight from Kolkata Airport. He must have heaved a sigh of

relief as he did so, as his trip was finally materializing after multiple false starts, bookings and re-bookings. The aircraft took a pit stop in Rangoon, Myanmar (currently known as Yangon), and at the airfield, the eminent ornithologist noted down observations of a Pied Harrier and the *insolens* subspecies of the House Crow, before the plane took off once again. Its flight path was directly overhead the Irrawaddy River basin and Abdulali was fortunate enough to get an aerial glimpse of the magnificent Golden Pagoda. His final destination, in his own words, was 'Green Islands in a sea of blue' – the singularly beautiful Andaman and Nicobar Islands.[1]

The first few days and nights he spent looking for birds, frogs and reptiles. His detailed notes contained information on how the Narrow-headed and Cricket Frogs, and the *Hemidactylus* geckos sounded different from their Bombay counterparts. His daytime exploits talked about being accompanied by an armed escort to protect him from the so-called dangerous Jarawas (even though the armed escort had never seen the Jarawa in the last five years!). Along with the birds he observed, he wrote about the disappearance of forest tracts around Port Blair and how invasive species, such as House Sparrows and Common Mynas, were making footholds in the disturbed habitat, while endemic species were losing out. He managed to get to Car Nicobar and even got lost while chasing Nicobar Imperial Pigeons (a feat impossible today, as the entire span of Car Nicobar is full of coconut groves, with hardly any native forest remaining).

However, the 1963 visit was just a reconnaissance trip – the first of eight that he made between 1963 and 1977[1-6] on behalf of the Bombay Natural History Society (BNHS). Abdulali was the first ornithologist to study these islands in some detail in nearly five decades, and the main aim of the expeditions was to collect bird specimens from these under-appreciated islands.

The taxonomy of the avifauna of the Andaman and Nicobar Islands received a lot of attention during the pre-Independence era. The earliest bird collections from the islands date back to the mid-1800s, and in parallel to what happened across the rest of the world, most bird species from these islands were described to science during the 'golden era of species discoveries', which ran from the mid-nineteenth century to the early twentieth century. (In order to formally describe a new bird species to science, a specimen is collected, so that the same can be used as a reference for all future studies.) Some of the notable early contributions came from E. Blyth,[7-11] R.C. Tytler,[12-13] V. Ball,[14-16] A.O. Hume[17-21] and A.L. Butler,[22-25] based on the collection of specimens made from the islands. Most of the Hume collection came from W.R. Davison, who was contracted by Hume to carry out the collection of specimens. The next series of species descriptions came from C.W. Richmond in 1902,[26] though the actual collections were made by Dr W.L. Abbott and C.B. Kloss. However, none of these collections were housed in India. This is what motivated Abdulali to visit these remote islands. He had always been interested in studying the taxonomy of birds and believed that was the way forward for BNHS.

Abdulali's field expeditions took him all over the Andaman and Nicobar archipelago, from the pyramid-shaped Narcondam Island to the southernmost Great Nicobar Island and everything in between. The outcome of these expeditions was nothing short of exceptional. He described no less than nine subspecies of birds from the islands alone, including Black Baza (subspecies *andamanica*) from Andaman, Slaty-breasted Rail (subspecies *nicobariensis*) from Nicobars, two subspecies of White-breasted Waterhens (subspecies *midnicobaricus* from Central Nicobar, and subspecies *leucocephalus* from Car Nicobar), Andaman Cuckoo-Dove (subspecies *tiwarii*), Brown Hawk Owl (subspecies *rexpimenti*) from Great Nicobar, Asian Fairy Bluebird (subspecies *andamanica*) from Andaman, white-eyed subspecies of Glossy Starling (subspecies *albiris*) from Nicobars and Black-hooded Oriole (subspecies *reubeni*) from the Andaman Islands.[1-6]

Abdulali's focus went beyond collecting specimens of birds and other taxa. He did his fair share for the conservation of wildlife. Thanks to his efforts, the Bombay Wild Birds and Wild Animals Act of 1951 came into being, curbing the rampant hunting that had begun soon after India's independence. He was also instrumental in getting nearly 105 sq. km of forest declared as a National Park just north of Aarey Milk Colony in Mumbai. It was called Borivali National Park then. Over the years, the national park (now called the Sanjay Gandhi National Park or SGNP) has been a mecca for every budding and experienced naturalist in the Mumbai region, serving as the port of embarkation for their journeys into the world of wildlife, myself included; and for that, we will forever remain in debt to Abdulali. He also played a crucial role in banning the export of frog legs during the monsoon months. In the 1960s, the number of Indian Bullfrogs (*Hoplobatrachus tigerinus*) had dwindled due to the commercial export of the species. Recognizing the critical role played by these amphibians in controlling insect populations, which helped agriculture commercially, Abdulali published a detailed report. This report eventually led to the protection of frogs under the Wildlife Protection Act.

Abdulali had other (non-wildlife) interests, too – he had a passion for driving motorbikes and cars – and was quite a personality; according to some of my friends, he was known for his sharp tongue as well. Working with BNHS for 70 odd years, he became a mainstay of the organization, and took the Society forward. He even managed to get permission from the Prince of Wales Museum to house BNHS on their premises, which is where the Hornbill House proudly stands today. The BNHS collection room, where he spent years of his life meticulously cataloguing every bird specimen at BNHS (more than 27,000+ skins), is named after him.

Three years after his first visit to the Andamans, Abdulali finally managed to visit the island of Great Nicobar. His word picture of Campbell Bay differs vastly from what we know of it now. He described the hamlet as a mere clearing, 'impenetrable on one side [due to the tropical jungle] and the sea on the other.' Working conditions were clearly not easy – his report notes that he lost about ten days of work, though it remains unclear whether it was due to a tropical illness or the impenetrable nature of the habitat – and his movements were somewhat restricted to the beach during low tide and a few hundred yards inside the forest, which was as much as he could penetrate through the tangle. On 3 March 1966, he collected a scops-owl which flew over this very clearing.[3] The bird looked very unfamiliar to him. There were no other records of scops-owls from Great Nicobar, and just two other records of scops-owls from the whole of the Nicobar chain – both had been collected from the island of Kamorta, in Central Nicobars, in 1873.[21] The male bird Abdulali collected from Campbell Bay had enlarged testes, which meant that the bird was in breeding condition. The bird had eaten a beetle and a spider, which he discovered during the taxidermy of the specimen.

A second female specimen of the same owl was collected by S.S. Saha of the Zoological Survey of India, on 2 April 1977 from Great Nicobar.[27] The specimen was accompanied by a drawing, which depicted a 'granular ovary', another indication that this too was a breeding individual. However, the individual had moulting inner primaries, and its stomach contained a mangled 4-in-long gecko. Abdulali likely believed this was the same kind of scops owl – what we know as the *Otus sunia nicobaricus*, a subspecies of the Oriental Scops Owl – that Hume had described from Kamorta in 1876.[21] It got the same treatment and nomenclature from Ali and Ripley in their handbook, which is widely regarded as the magnum opus of Indian ornithology.[28]

However, it was American ornithologist Dr Joe T. Marshall[3] who noticed that the wings of the scops-owl from Great Nicobar were too large for it to be any other species of scops-owl – this observation was possibly made during his visit to the BNHS. However, when he examined the BNHS specimen, he was unable to identify it as any known species of scops owl. He wished to visit Great Nicobar (possibly to confirm this), but by then the Great Nicobar Islands were out of bounds for all foreigners, and his questions would remain unanswered.

As the years passed, these specimens of scops-owls from the Nicobars lay somewhat forgotten in the drawers of the BNHS collection. They were referenced, though, in a paper by Ripley and Beeler on the region's biogeography,[29] as well as in one written by the Indian ornithologist Ravi Sankaran.[30] However, three decades after Abdulali collected the first specimen in 1966, they drew the attention of American ornithologist Pamela C. Rasmussen, then working with the Smithsonian Institution in Washington DC, United States. Within the space of a week, she examined the Kamorta specimens preserved at the Natural History Museum, Tring, in the United Kingdom, and the ones at BNHS, and realized that the two sets differed significantly. Rassmussen went on to formally describe the Nicobar Scops Owl as a new species, *Otus alius*, in 1998,[27] naming it after Abdulali's family name.

After its formal description by Rassmussen, the species made two appearances – the first appearance was in the form of sonograms – two kinds of the owl's songs were recorded by Indian Forest Service officer Pratap Singh. However, these recordings are unavailable in the public domain. In the second instance, there was a photograph from 2014, when a local bird guide from Andaman, Vikram Shil, and bird photographer Jainy K. bumped into the species, also in Campbell Bay. But reliable sightings of the bird remained elusive, even though the island was visited by a handful of other birders.

In 2015 I was on a mission to see as many bird species as possible within India in a single calendar year, and it was impossible to complete the 'Big Year' without visiting the Nicobar Islands. A Nicobar birding trip was in fact a long-cherished dream of mine. I had come close to visiting Great Nicobar in 2010 with a very close friend, the late Ramki Sreenivasan; all our plans were in place, and we had even obtained the coveted Tribal Pass, without which a trip to Central Nicobar is impossible. That plan didn't pan out due to personal reasons, but thanks to it, I had already done my homework for a Great Nicobar trip, and also made a list of birds I wanted to see. The list was short but full of mega-species and subspecies, many of which were species or populations that no one had seen for years. And during my Big Year, I decided to make a determined effort to get there.

People often speak of the Andaman and Nicobar Islands in one breath, without even thinking about or understanding the islands' geography. But this is supremely important in order to grasp the logistics involved in getting to the Great Nicobar Islands. The Andaman and Nicobar Islands are part of a long linear chain, with North, Middle and South Andaman mostly being well-connected. The capital city of Port Blair is located in South Andaman and houses a third of the island's population. Little Andaman is located just south of South Andaman, and is separated from it by a shallow strait, called the Duncan Passage. The town of Hut Bay is the most important location on Little Andaman. Then come the infamous rough seas of the Ten Degree Channel, which separates the Andamans from Car Nicobar. As you head further south, another deep-water channel separates Car Nicobar from the Central Nicobar Islands. Finally, further south of Central Nicobar Islands is another deep channel, called the Sombrero Channel, and crossing it brings you to the islands of Little and Great Nicobar. In effect, the long Andaman and Nicobar chain abuts Myanmar in the north and extends all the way to approach Sumatra in the south.

I landed in Port Blair on 21 March 2015, and the first five days were spent 'cleaning up' all the Andaman endemics. My wife Vishnupriya then joined me, and we decided to head to Campbell Bay, located 529 km from Port Blair. One way of getting there is by ship. Although three or four ships plied the route, their schedule was erratic and infrequent, and I had little time to squander during an action-packed

Big Year. The second and more exciting option was to take a Pawan Hans–operated helicopter, which avoided travel delays. However, this was much easier said than done. It took a while and a lot of convincing the local authorities before Vishnu and I managed to secure two seats on the helicopter from Port Blair to Campbell Bay, albeit on two separate days. I would fly to Campbell Bay first and Vishnu would follow me a day later.

I was euphoric about heading to Nicobar and couldn't catch a wink of sleep all night. This was quite understandable as, just a few hours earlier I wasn't even sure whether I would ever get to Great Nicobar, let alone the idea that I would fly in a helicopter to the place of my dreams. For years, I had been chatting about my Nicobar birding plans with one of my close friends, James Eaton, because, quite simply, that's what birders do. Until they get to their dream birding destination, they talk about it, and sometimes quite obsessively. They discuss the birds, strategies and logistics, among other things. James and I had discussed the Nicobars several times earlier. Even though Great Nicobar was open for Indians, in 2015 (and even now), the whole of Nicobars was out of bounds for foreign birders; meanwhile, Central Nicobar and Car Nicobar were only accessible to Indians who managed to get their hands on the Tribal Pass. James had explored a group of tiny islands just off Sumatra, called the Barusan Islands. These islands, like Simeulue, Nias, Siberut, Sipora and Pagai, are biogeographically contiguous with the Nicobar Islands. And so, James had great insights regarding the birds of the Barusan islands.

Hours before my helicopter departure, while I was struggling to find Internet connectivity in Port Blair, James was leading a bird tour in Vietnam with slightly better connectivity. He was watching the endemic Sooty Babbler there when I pinged him with a text about my Nicobar plans. He shared my enthusiasm for the trip (though it was tinged with a bit of good-natured jealousy!), and we continued chatting and exchanging various notes. Then it suddenly struck me – I had read in Rasmussen's paper[25] that the 'wing formula' of the Simeulue Scops Owl and Enggano Scops Owl (another species James was familiar with) and Nicobar Scops Owl is very similar. So, I requested him to share his recordings of the Simeulue Scops Owl. James managed to take time out from his hectic tour schedule to send me three recordings of this tiny Scops Owl, which is endemic to Simeulue Island. With a near-miraculous spike in connectivity (which sounds like fiction in the always-connected universe of 2024), I downloaded those calls on my cell phone just in time. The logic was simple – the island of Simeulue is closest to Great Nicobar and equally isolated from the mainland (Sumatra) by deep seas; there was definitely a possibility that the Nicobar Scops Owl was related to the Simeulue Scops Owl.

The excitement of getting to Nicobar continued to keep me awake, till it was time to head to the airport, armed only with my binoculars, camera, sound recording equipment (including a massive parabola), toothbrush and toothpaste. I couldn't carry anything else as the helicopter had a strict baggage allowance limit of 5 kg per person. Trusting Vishnu to get some clothes for me the next day (the hope being that the extra weight would be offset by her extremely slight – and luckily light – body mass), I left for the helicopter hangar at the crack of dawn.

The journey was truly spectacular – the three-hour, 500 km-plus flight presents the traveller with breathtaking views of tropical islands filled with majestic rainforests, mangroves and blue seas studded with coral reefs. On deplaning at Campbell Bay, the skin instantly soaked in the hot and humid equatorial weather. But birding in Nicobar proved quite easy. A short evening walk in the vicinity of zero-point at Campbell Bay produced many of the common Pied Imperial Pigeons, some Chinese Pond Herons and Chinese Sparrowhawks. A stroll into the very first small forest patch produced a Nicobar Hooded Pitta, in just two minutes. I later realized that this bird was common all over the island. A great view of the loud, noisy endemic Nicobar Parakeets followed this excellent start.

As dusk approached, I found a suitable habitat, which was between a forest patch on one side and a degraded area on the other. I then played one of James's recordings of the Simeulue Scops Owl. The response was immediate – an owl flew out of the forest patch and landed on a bendy bamboo stem; its yellow eyes, piercing through the darkness, fixed on me. Adrenaline ran thick and fast through my body. It was 'my kind of birding' moment. A combination of homework, hard work, overcoming the difficulties of getting to Great Nicobar, a good prediction of habitat and timing had culminated in the sighting of this rarely seen species (or so I thought at that moment.). The bird was dark and heavily barred. The eyes were yellow with black edges, and the eye rings pink, giving it an intense, scary and comical look all at the same time. I started to walk back to the guest house. On my way, I found four more individuals of the Nicobar Scops Owls before I turned in for the night. What an exciting and action-packed first day in Great Nicobar!

The following day, Vishnu joined me in my birding quest, and we spent the next week birding on the only two roads built on that island (at least we could move around, unlike Abdulali in 1966). Now familiar with the Nicobar Scops Owls, we also came across a bold pair of Brown Hawk Owls (endemic subspecies: *rexpimenti*). Further excursions to the local forested areas produced one of the most awaited birds for both of us – a Nicobar Jungle Flycatcher. The species first became known to the ornithological world in 1903, when American ornithologist C.W. Richmond[31] wrote about how 'Abbott and Kloss found it to be common' in Great Nicobar. It then vanished from all knowledge till it was re-found by Abdulali in 1966. After this, the flycatcher made two other appearances – a song recording by Pratap Singh, and a photograph by A.P. Zabin from Little Nicobar. We were thrilled to sight such a rare bird but soon found them singing in many regions of Great Nicobar – which brings to mind the maxim that a bird is rare only until you find it!

We then 'twitched' all other Great Nicobar endemics (species as well as subspecies), such as the Great Nicobar Serpent Eagle, Nicobar Imperial Pigeon, Crimson Sunbird (endemic subspecies: *nicobarica*), Olive-backed Sunbird (endemic subspecies: *klossi*), Indian White-eye (endemic subspecies: *nicobaricus*), Oriental Dwarf Kingfisher (subspecies: *macrocarpus*). All these subspecies look quite distinct from mainland populations and also from those on the other islands of the Andamans and Central Nicobars. We also managed to spot two (then) vagrant species for the Indian subcontinent – the Arctic Warbler and Daurian/Purple-backed Starling. Though unknown to Indian birders then, both species can be regularly sighted in the Andaman and Nicobar Islands. We also came across the Great Nicobar Crake, a bird that still requires a formal description. The only endemic that I had an unsatisfactory sighting of was the Nicobar Megapode, so I refused to count it on my 'life list'. I had to wait another eight and a half months to count one in feathers.

I will always cherish my very first visit to the Nicobars, not only for the birds but also for the tiny sea turtle hatchlings (Olive Ridley and Green), the squeaky Nicobar Tree Shrews, the dark-coated Crab-eating Macaques and, of course, the spectacular vistas of untouched primary forest, and seas so blue that the camera couldn't capture all its shades.

It didn't take me long to visit the Nicobars again; eight and a half months later, in December 2015, I found myself back in Great Nicobar, and managed to see all its bird endemics this time, including the Nicobar Megapode. On the same trip, I also managed to get to Kamorta in the Central Nicobars, where I found and sound recorded the rufous-coloured scops owl, which is often referred to as a subspecies of the Oriental Scops Owl, *Otus sunia nicobaricus*. Those were the first sound recordings and just the third sighting ever of this bird. However, that's another thrilling tale for another time.

Since 2015, I have visited all of the Nicobar Islands multiple times and come to know the Nicobar Scops Owl a little better. I know now that the Nicobar Scops Owl is not just restricted to Great Nicobar but can also be found on the nearby Little Nicobar Island. I know they have three different song types, and they start breeding sometime in February. The breeding continues until April, and by May the fully feathered juveniles fly around, but are still fed by the adults. The juveniles are less dark than the adults and much whiter on the belly, and have a distinct single note insect-like contact/begging call. Post breeding, adults finish moulting their outer primaries by the end of May. From the population point of view, the species is widespread on the island and can be found both in degraded habitats and densely forested areas. Even though I have seen these tiny owls hundreds of times now, I still get excited about seeing one every time I step out for an owling session.

So, what does one need to do to see this species in the wild? Like most of the other species found on remote islands, one needs to tackle the logistics first. Once you finally make it to Campbell Bay, it may not take more than

15 minutes after dark to find one. As you watch it, elated at bagging a 'lifer' so fast, I am sure you will find it staring right back at you, with yellow gleaming eyes and pink eye rings!

7

In Search of the Last Megapodes

RADHIKA RAJ

The Nicobar Megapode is an iconic and endemic species. Highly localized, you can only find the bird within parts of the Nicobar chain of the Andaman and Nicobar Island group. Though found in several Nicobar Islands, a major part of the bird's population lives along a narrow coastal belt. As Radhika Raj evocatively captures in her essay, this makes the species vulnerable to disasters, such as the 2004 Indian Ocean Tsunami that reset the shorelines of these islands; as she points out, this one single event caused the bird's population to plummet by about 70 per cent. Not just the Tsunami, the islands are vulnerable to every kind of stochastic event, including climate change. Which is why we need to grab every opportunity to conserve the bird across its habitat rather than threaten that habitat in any way.

RARE FEATHERS

You've read about them, now take in the beauty of India's rarest feathered treasures. Each image tells the story of a unique habitat, and the passionate quest to see these species.

A Pink-headed Duck pair, painted by the Danish artist, Henrik Gronvold (1858–1940), in 1908 (Stuart Baker, 1908, *The Indian Ducks and their Allies*. 2nd ed.).
Photo credit: Wikipedia

Mount Victoria Babax at Phawngpui, Mizoram
Photo credit: Andrew Spencer

That piercing yellow gaze! A Nicobar Scops Owl from Great Nicobar.
Photo credit: Shashank Dalvi

Chin Hills Wren Babbler, a species restricted to the Chin Hills Mountain Range in Myanmar and Mizoram in India.
Photo credit: Shashank Dalvi

Hodgson's Frogmouth: note the hair-like feathers on its head that protect the bird from insects.
Photo credit: Shashank Dalvi

Banasura Laughingthrush – well worth the climb to its montane habitat.

Photo credit: Shashank Dalvi

A largely diurnal species, the Forest Owlet, photographed at Tansa Wildlife Sanctuary, just three hours from central Mumbai.

Photo credit: Shashank Dalvi

A Nicobar Megapode on its mound in Great Nicobar.

Photo credit: Dhritiman Mukherjee/Roundglass Sustain

A Masked Finfoot in its habitat in the Bangladesh Sundarbans.
Photo credit: James Eaton

A male Mrs Hume's Pheasant at its night roost.
Photo credit: Dhritiman Mukherjee

A singing Long-billed Bush Warbler perched on Rumax, in what may the last few patches of its habitat in the Naltar Valley of Gilgit-Baltistan.

Photo credit: James Eaton

'I can tell you where exactly to look for the Nicobar Megapode, but I can't guarantee you will see it,' says Shashank Dalvi, one of India's leading ornithologists. A few hours after we land in Great Nicobar – the last island of the Andaman and Nicobar archipelago – we bump into Dalvi at a local tea stall. Dalvi, at 40, is a human dynamo with the enthusiasm of a student despite his greying hair. True to his character, he wears a mask that says 'Bird Nerd' (a gift from his wife), while a Sri Lankan Frogmouth, a frog-faced bird, stares at us from his worn-out T-shirt.

It is March 2021. The first wave of the deadly COVID-19 pandemic is waning – cases are falling, offering a brief respite. The world, cautiously, has opened up. Dalvi has moved into a garish pink room above a *dhaba* in Campbell Bay, the island's only town, along with two assistants, to study the birds of the island. Wildlife photographer Dhritiman Mukherjee, a crew of filmmakers and I are here to document the rare species of the archipelago. The elusive Nicobar Megapode (*Megapodius nicobariensis*), an endangered bird found nowhere else in the world but the Nicobar group of islands, is on both our lists.[1] 'It's pretty simple,' says Dalvi. 'Look for a huge dead tree. Check if a large mud mound circles its base. If you spot it, move away and settle down. It could take a while before the bird makes an appearance.'

The Andaman and Nicobar archipelago is a crescent-shaped trail of over 500 islands located in the Bay of Bengal, separated from mainland India by about 1,200 km. The Great Nicobar Island is the last of the lot. It has taken us a flight from Delhi to Port Blair and a two-day journey on a cargo-cum-passenger ship which runs on an unreliable schedule, to reach here. From here, Indonesia (180 odd km away) is closer than Kanyakumari. The island's azure waters and silver beaches resemble photographs of Southeast Asia's top travel destinations, but there are no tourists or hotels in sight. Instead, 80 per cent of the Island is demarcated as a biosphere reserve that hosts rich and threatened tropical rainforests.

In 1969, 330 retired servicemen and their families were settled here and awarded 12–20 acres of land each for their move. The scheme – initiated by Indira Gandhi's government – was a strategic decision to populate the country's isolated southern tip. The first batch of families from Punjab arrived on a navy vessel with groceries, clothes and mosquito nets. They cleared large tracts of forest, built homes and cultivated farms from nothing. Today, that population has grown to barely 8,500 individuals who live on a narrow strip along the island's eastern coast. Apart from them, government officials and the indigenous tribes that live in harmony with its forests, Great Nicobar is sparingly populated. Few outsiders venture this far. However, every once in a while, around a tea stall, you spot a motley crew with backpacks, notepads and binoculars hanging off their shoulders, all desperately in need of a shower.

After all, remote islands are irresistible to ecologists. Remarkable things happen here. Isolated by the ocean, disconnected from the mainland, species here evolve independently – and radically differently – into 'endemic' versions that exist nowhere else in the world. Islands have witnessed the rise and fall of the oddest creatures – from oversized rodents and pygmy mammoths to giant flightless birds. Unfortunately, what makes endemic species special, also makes them acutely vulnerable.

A single epidemic or a minor calamity can send an entire geographically restricted population over the edge. Add human-led destruction to the mix and you have a recipe for disaster. Since the 1500s, more than 75 per cent of known extinctions have happened on islands. New research claims that the likelihood of an insular or endemic island species being driven to extinction by humans is twelve times higher than that of a continental one.[2]

The archipelago we've travelled to is just as vulnerable. Of more than 9,100 species present in the Andaman and Nicobar archipelago, a staggering 1,032 – including the Nicobar Megapode – are endemic.

Named after their monstrous feet – the word 'megapode' literally means 'big foot' – the birds are not much to look at. Twenty-two megapode species occur on islands in the Australasian region, such as the western Pacific, Australia, New Guinea, Indonesia and Nicobar groups.[3] Most resemble large chickens or stocky turkeys.[4] The Nicobar Megapode looks like a big brown chicken with a blood-red head.

However, one thing remains common to all megapodes – none use their bodies to incubate eggs. Instead, they conjure up ingenious strategies to harness external heat to hatch them. The Melanesian Megapodes, found exclusively in the

Solomon Islands, lay eggs deep in volcanic ash and depend on the warmth of a live subterranean volcano to incubate them. The Nicobar Megapodes, the ones we are desperately hoping to see, use their disproportionately large feet to build enormous mounds near the coast with sand, corals, shells and rotting vegetation. At the heart of the mound, in a burrow lined with dry leaves, they lay salmon-pink eggs. As the vegetation rots, it produces enough heat to incubate them. Built and rebuilt on by multiple pairs, year after year, the nests of these chicken-like birds can tower over a standing human and run as wide as a tennis court.

For hundreds of years, these mega mounds have amazed everyone who has stumbled upon them.

The oldest written records of these mound nests in India date back to the eighteenth century, when European naturalists travelled to explore the riches of the islands during the colonial regime. One of the most descriptive accounts of these unusual birds and their nests was penned by A.O. Hume, a British civil servant and ornithologist known for his extensive documentation of the birds on the subcontinent (though he is more famously known for being the founding member of the Indian National Congress, the party that eventually led India to Independence). In 1874, during a month-long trip to the island, he spotted several megapodes and mounds, especially on Galathea Bay, along the island's eastern coast. 'They are to be met with in pairs, coveys, and flocks of from thirty to fifty. They run with great rapidity and rise unwillingly, running and flying just like jungle hens,' he wrote in *The Game Birds of India, Burmah, and Ceylon*. However, it was the giant nests that left him astonished. 'Moderate-sized birds as they are, they gradually manage to accumulate tumuli [mound of earth raised over a grave] that would not have done discredit to the final resting place of some ancient British hero,' he wrote.[5]

Over a century later, we are headed to the same coast where Hume spotted several large flocks of the Nicobar Megapode.

As we take the road south to Galathea Bay, we quickly realize that locations along the coast are marked by the number of kilometres they are from 'Zero Point', a small market at the heart of Campbell Bay. The island's vast uninhabited stretches have inspired few tales or names. Today, we are going to '41', where the forest department's camp is located. At some point, we cross the last school, and then the last police station. By now, we have left all worldly comforts – electricity, running water, phone networks, toilets – far behind. A small shop called the 'Southernmost General Store of India' once marked the 'end of civilization', but it was swept away by a major natural disaster, one of the worst in the region's history. On the morning of 26 December 2004, just as Christmas festivities wound down, a deadly tsunami – triggered by an undersea earthquake in the Indian Ocean about 125 km south-west of Great Nicobar – slammed into the island's coast. The coastlines of Great Nicobar went significantly under, as much as 4 m at the southern tip, causing sea waters to sweep inland. Along the road that hugs the coast, naked stumps of trees stand as reminders of the monstrous disaster.

We get off the road at '41' but our destination is still an obstacle course away.[6] First, in the fading light, we wade through the hip-deep waters of a small canal, which I only later find out is a haunt for saltwater crocodiles. Next, we hike through a squelching mangrove forest to reach the banks of the Galathea River, where Jugloo and Prem – forest guards and the island's secret keepers – have been waiting for us. They paddle us across the river's eerily quiet waters on a bamboo raft, bound by twine and flanked by blocks of thermocol for balance and buoyancy. We soon reach the edge of a dark forest. Jugloo and Prem, spry and scorched by the sun, have grown up on the islands and know these forests intimately. They move through it like shadows, walking unerringly through the heart of the jungle, where all paths disappear. We follow them quietly until we finally reach a narrow strip of shore that faces an endless ocean.

A wooden shack set up by the forest department which smells of pungent fish curry, welcomes us. We dump our luggage and set up tents under the stars. The moon has given us a miss; instead, a blazing Milky Way rises from the sea and disappears into the rainforest behind us. Barefoot, I step on the wet sand, and it lights up like a firecracker; the sparks are caused by marine plankton that emit light when disturbed. I look up and a shooting star cuts through a constellation. Nothing has prepared me for Galathea. My head swirls, my knees are weak. But there is no time to take this all in; as dusk falls, swarms of sandflies, notorious for colonizing the most pristine beaches, attack, and I feel like a thousand needles are stinging my body. Pests! Galathea Bay is also one of the largest nesting sites in the country and the Northern Indian Ocean, for the Leatherback – the world's largest sea turtle. All night, we take turns to scan the shore, eagerly waiting for a giant female to crawl out of the waters. Sleep will have to wait.

The following day, just before dawn, the megapodes sing. The female leads the duet with low-pitched wails. As the male joins in, they reach a crescendo that ends with a loud, shrill, metallic clattering. Jugloo and Prem rush us into the mangrove thickets that line the coast. Beyond the dense veil, a large, dead tree stands circled by a 3 ft mound. It looks like a shrine studded with twigs, seashells, plastic bottles and pop-coloured rubber slippers that have floated in with the tide. Dead wood. Check. A mud mound. Check. It is time to build a hide.

Building a hide, far removed from civilization, demands innovation. The Nicobar Megapode is a skittish bird, known to flee at the slightest suspicion. Our success at spotting it will depend on how well we can blend. Mukherjee directs the construction like the conductor of an orchestra. Dead logs of wood are scavenged and tied together with jungle vines. A large coconut palm leaf makes for a great front cover, with natural gaps to stick camera lenses through. The four of us squeeze in, elbows and knees sticking into each other. Five eventless hours pass. We run out of water. A

Monitor Lizard, known to feast on megapode eggs, surveys the burrow. But no megapode.

Our first attempt gives us a taste of what the next three weeks will be like – lots of waiting, sweating and nothing. Jugloo and Prem reassure us. They know the megapode. As children, they hunted the *jungli murgi*. But in the 1990s, the late conservation biologist Dr Ravi Sankaran and his student K. Sivakumar employed their tracking skills to protect the endangered bird. Over six years, Jugloo accompanied them across Nicobar's sixteen islands, covering over 600 km, mostly on foot. Almost everything we know about the megapode comes from their work.

Dr Ravi Sankaran is something of a legend in ecological circles, especially for his work on birds that are found in the most challenging landscapes that nobody bothered to study. But also because he was known to push boundaries, often at great personal risk, and encouraged others to do so. Fondly called the Indiana Jones of Indian conservation, Sankaran worked in the most primitive conditions for eight years in the Andaman and Nicobar Islands to document its endemics. On these islands, he remains a memorable figure. The Nicobarese from Chingen Village, a settlement along the coast of Great Nicobar that remain deeply suspicious of mainlanders, fondly referred to him as *Chidiya Babu*. Under his supervision, Sivakumar went on to pursue the first-ever doctoral study on the Nicobar Megapode which, two decades later, is the only detailed work on the bird.

Later, when I return to Delhi, I watch Sivakumar's online lectures and speak to him over a Zoom call. Dressed in a clean-pressed, buttoned-up blue shirt, thick-rimmed glasses, tamed hair and a trimmed moustache, Sivakumar comes across as someone who enjoys the comforts of a routine and an office. Fortunately, by now, I know better – ornithologists are closet adrenalin junkies. 'Dr Sankaran was ready to risk his life for birds,' he says. 'I couldn't have found a better guru.'

Sivakumar had grown up in a small town near Pondicherry and secured a degree in microbiology from a local college. He had never lived away from home, or spoken any language other than Tamil and English, and knew nothing about wildlife. And yet, a job advertisement in a local newspaper looking for someone to work on endemic birds in the remote Nicobar Islands caught his fancy. 'I was in search of a challenge, something exciting,' he recalls. 'I wanted to travel the furthest, do something nobody had done before.' Sankaran, who was known to be a fine judge of character, had found his match.

The year was 1995. On the very first day, after they landed in Campbell Bay, Sankaran and Sivakumar walked for 35 km or 14 hours, with ration and luggage on their backs, along a damaged road to reach Galathea Bay. 'For hours, I did not see a single human being. I remember spotting coconut trees and thinking that if I don't find food, I might be able to survive on those,' he says. For the first four nights, while they constructed a 'camp' – a straw roof, suspended on areca palm wood poles – Sankaran, Sivakumar and Jugloo slept on the open beach, under mosquito nets that were a flimsy defense against the monstrous sandflies.

On the eleventh day of their visit, Sankaran came down with malaria and walked 35 km back to admit himself to the only hospital on Great Nicobar in Campbell Bay. Sivakumar was left alone with a much younger Jugloo, a first-time field assistant with a penchant for disappearing for weeks, especially when he got his hands on *undiya*, a tribal rice beer. And yet, there was something that made him stay, perhaps the enigma of the megapode – 'I couldn't believe that such a small bird could build a nest larger than me!'

The mound nest is the core of the megapode universe. A solitary male must charm a female before the pair can embark on building a nest or securing a spot on an already existing mound. The construction involves the careful selection of materials that are mostly found only in very specific habitats – where deciduous forests line the shore. Clay-like soil that is neither sandy nor loamy is mixed with corals, shells, rocks and leaf litter to construct a mound.

Around four pairs use the same mound and form overlapping territories around it. The picture becomes clearer as I see some of Sivakumar's sketches. ('Dr Sankaran had advised me to not carry a camera to the field. There was no need for unnecessary distractions,' remembers Sivakumar, who turned to sketching the bird and its nest mound.) With the mound at the centre and a quadrilateral for every megapode territory that intersects at it, the diagram looks like an asymmetrical, lopsided pinwheel. During the breeding season (February–May), the site turns into an arena. Megapodes are highly territorial. Twice a day, every morning and evening, the pairs show up to stake their claim. The male climbs the mound and sounds a territorial call – a chuckle and a croak. 'As if to say, "I am here, this is my mound",' says Sivakumar. There is an unspoken agreement among the couples – one doesn't show up when the other is there. On most days, a call is enough to ward off intruders, but on rare occasions, a challenger shows up. Fights erupt. Solitary males chase females to win them over, and males battle to capture the fort. I see Sivakumar's hand-drawn sketches of megapode fights. It's all legwork – the first line of attack targets the feet, then a flying kick to the chest, and if matters get out of hand, a severe blow on the head settles the duel.

Once the pairs are committed to each other, they mate, and the female lays a clutch of 2–3 eggs. For at least seven weeks, the megapode pairs regularly visit their burrows to check on the eggs. They inspect the mound to keep the temperature at about 33°C, perfect for incubation. If the temperature drops, they add more leaf litter to the mix. If the temperature rises, they dig pits to release the heat. Their close monitoring of the mound's heat gives the megapode family the name 'temperature birds'. About seven weeks later, the eggs hatch and the chicks dig themselves out of the mound to emerge in the most mature condition of any bird. They can run, fly and even hunt on the same day. No parenting, or parents, required.

For three and a half years, Sivakumar was privy to this exceptional avian drama. Until Sivakumar's research, the Nicobar Megapodes were assumed to be strictly monogamous, but he saw pairs split, find new partners and then reunite several breeding seasons later.[7] On some days, things got ugly. On one occasion a male tried to forcibly mate with the female from another pair. The chase and fight lasted for 45 minutes, while their respective partners stood, calling incessantly in vain. 'Even in the mid-90s, the megapode's habitat was being taken over by locals for agriculture, leaving the birds little room to build mounds. Without a mound a male has no chance to win over a female. Perhaps, this was the reason megapodes were trying to mate with already paired females,' says Sivakumar.

Fieldwork often wrapped up by noon, and Sivakumar spent the rest of his day watching the rarest creatures only found on the Nicobar Islands – Robber Crabs, Nicobar Treeshrews, Great Nicobar Parakeets and sea turtles. Many kept him company. At some point, a snake moved into Sivakumar's open hut, and he tiptoed around it for weeks. More than once, a Green Sea Turtle chose to build a nest on the bare floor of his hut. 'Galathea is the closest thing to paradise,' he says.

But surviving paradise needs reckless courage that flirts with death. The first time Sivakumar got malaria, he walked 35 km to Campbell Bay, like his mentor had done within two weeks of their arrival, and admitted himself to the same hospital. Over three and a half years, he'd have to do it six more times. 'You reach the hospital, rest, have ample coconut water and return to the camp,' he says casually of his malaria cure drill.

As years passed, things improved. The forest department lent him a tarpaulin sheet to cover the straw hut. Sivakumar arranged for a kerosene lamp, with fuel just enough to light up the hut for 30 minutes every night. He made new friends in Chingen and learned to roast freshly caught fish over a wood fire.

Sivakumar's grit is aspirational. We've been on Galathea for over two weeks, and the megapode has eluded us. I have tried every trick to keep the sandflies away, but my arms are pockmarked with pustules. During a trip back to Campbell Bay to charge equipment and take a much-needed shower, we reconnect with Dalvi. Mukherjee and Dalvi, old friends who have travelled together to the remotest places, are unperturbed by the sandfly quandary. Instead, they list insects with the fiercest bites. 'In the summer, ticks in Bhadra and Kudremukh in Karnataka are deadly,' says Dalvi, but Mukherjee argues that his worst nemesis is the *pisu*, which he encountered in the rainforests of Northeastern India. The sandfly, which has kept me up for nights, fares poorly.

Like adventures often do, Galathea leaves me awestruck and awfully exhausted. On some days, we hear the megapode at a distance, but we cannot see it. We turn to Jugloo and Prem for hope and advice. 'At noon it wanders in a forest at a distance from the mound,' says Prem. We climb a hill and build a hide under giant fig trees with emerald crowns and silver barks. The hours pass slowly. Fig seeds shaped like paragliders dance with the wind and gently land in my lap. The megapode, however, is a no-show. 'At night, the bird is brave,' says Jugloo. 'It flies to the tree branches and calls – a way to demarcate territory.' We follow the duets in the dark, torches in hand, but in vain.

Finally, disheartened, we decide to move locations. Prem guides us through the Galathea River, over a hill and through a canal to set up camp on the other end of the Bay. At 5 a.m., the megapode calls suspiciously close to our tents. Barely awake, Mukherjee grabs his camera and darts into the forest, barefoot. Minutes later, he is back, laughing. Ancient forests surround us, and he has a shot of the megapode running across the camp's dry toilet.

I am slowly beginning to lose hope, but Mukherjee, one of the country's most celebrated photographers, and a philosopher in disguise, is always alert. On a particularly slow day, I ask him what keeps him going. '*Mazza aata hai*. It's fun,' he says with a wicked grin. Drawn to extremities, Mukherjee has spent years in search of the rarest species in the country and has arguably shot the first professional photographs of many species, including the Western Tragopan and, more recently, the world's largest cave-dwelling blind fish, *Neolissochilus pnar*. 'Looking for a bird or animal sometimes feels like a puzzle that tests me physically and mentally, but it is also the only thing that truly helps me understand its behaviour, habitat and plight.

To search is to learn,' he says. 'We are in the megapode's habitat, why can't we see it? Our struggle is a sign.'

Every story on these islands is shaped by the tsunami. Sparked by the most powerful earthquake ever recorded in Asia, the Indian Ocean Tsunami killed over 2 lakh people across fourteen countries. Closer home, the archipelago, particularly the southern Nicobar Islands, were the worst hit. Of the 3,500 people reported missing or dead on the islands, at least 3,000 were from the Nicobar group.[8] Galathea, one of the closest points to the epicentre, saw waves as tall as six-storey buildings batter the coast. Ninety per cent of the mangrove forests were killed instantly. The tsunami was merciless, but the earthquake caused an irreversible seismic shift. The entire archipelago tilted. The edges of islands in the Andaman group were lifted out of the waters, exposing corals that died instantly. The coasts along the southern Nicobar Islands went under water. A few islands were submerged. The uninhabited Megapode Island and Wildlife Sanctuary, a small 0.2 sq. km islet west of Great Nicobar, was swallowed in its entirety. The island, named after the bird, had a small megapode population. Today, two large dead trees stick out of the waters as witnesses to the destruction.

In 2005, when Sivakumar and Sankaran travelled back to a tsunami-ridden island for a survey, they noticed that their field site was underwater. Ninety per cent of the ground-nesting bird's mound nests are built 30 m from the shore, just above the high-tide line. In 1994, Sivakumar had counted at least 35 active mounds along Galathea Bay. Most were washed away.[9] A single event had plummeted the bird's population by about 70 per cent.[10] 'I could only cry. There was nothing left, not a single person I knew, or a tree I had seen,' he said. It was also the last time Sivakumar would meet his *guru*. In 2009, the bird lost its most vocal champion. Sankaran, who had been instrumental in its conservation, died of a sudden heart attack near his home in Coimbatore.

Sivakumar's highly appreciated doctoral thesis secured him a job at the Wildlife Institute of India, an institution under the Ministry of Environment, Forest and Climate Change. Throughout his term, he tried directing funds towards the conservation of the Nicobar Megapode, but with little success.

In 2012, when former Prime Minister Manmohan Singh was to inaugurate The Eleventh Conference of Parties (COP 11) of the Convention on Biodiversity (CBD) held in Hyderabad, Sivakumar recommended that the megapode be featured on an Indian postal stamp as a part of a series that was released at the event. Today, the Rs 5 stamp is a collectible. Sivakumar is now a professor in the Department of Ecology and Environmental Sciences at Pondicherry University. 'The megapode gave me my career, but I am not sure if I can save the bird,' he says in a moment of rare vulnerability. 'The species is going down in front of me and I am helpless.'

The story of the megapode is also the story of the Nicobar group of islands; both exist on the margins of human imagination – ignored and dispensable.[11] The colonists used the islands to set up penal colonies, isolate political prisoners, and logged its rainforests indiscriminately for resources. Post-Independence, researchers have found that several islands from the Nicobar group, including Trak and Treis, have been used as targets for testing army ammunition.[12] In 2011, the Indian Navy proposed a missile-testing range on uninhabited Tillangchong, a narrow 17 sq. km island. The Navy had sought permission to use the forest land for the erection of a temporary structure as target for testing the accuracy of missiles fired from submarines. The news sparked protests among both tribals and ecologists. The Nicobarese consider the island sacred, an abode for the spirits of their ancestors. Hunting was declared taboo, too. A panel of experts and local leaders raised objections, and the plan was put on hold.

However, the most alarming development yet was evolving as we were traversing the coast, desperately looking for the bird. In January 2021, the government denotified the Galathea Bay Wildlife Sanctuary. The move is a part of an ambitious Rs 72,000 crore plan to build a transhipment port, an airport, a township and a power plant on Great Nicobar. About 70 per cent of the project sits on forest land, while two of its largest developments, the port and the airport, will spread across Galathea Bay. The environment ministry has approved the hacking of over 8 lakh trees for the construction.

Unfortunately, the mitigation measures seem perfunctory and insincere. The compensatory afforestation for the trees to be hacked is proposed to be done in the Aravalli hills in Haryana, 2,400 kms away from Great Nicobar. The government has also proposed the creation of three new wildlife sanctuaries in the archipelago, including a Megapode Sanctuary on Menchal, a 1.29 sq. km island overrun by coconut plantations.[13] A case of too little – way too little – for a project whose economic and ecological rationale is questionable.

The radical transformation of Great Nicobar Island may not only destroy the nesting site of the Megapodes and Leatherback Turtles but, activists insist, may severely impact the lives of two tribal groups – the Shompen and the Nicobarese. The particularly vulnerable Shompen, a hunter-gatherer community of about 200–300 individuals, have lived only in Great Nicobar for at least 2,000 years, with little contact with the outside world.

We recognize we are at the cusp of one of the largest transformations Great Nicobar has ever seen, and yet we hopelessly persist. It's been three weeks. We've built eight hides and hopped three coasts, and all we have are fleeting flashes of the megapode. On our last day, at 2.30 a.m., we get on a motorboat to head to Navidera, a coast north of Campbell Bay. As we get off the boat, the sun peeks from the shimmering horizon. We step on the shore and walk straight into the littoral forest and soon spot a dead tree and a mound that towers over me. It is the largest mound we have seen yet. We settle into a hide at about 5 a.m., taking turns to nap all day. Finally, at 4.30 p.m., we hear the familiar clattering song. And then, through the gaps in the palms fronds I see a blood-red head, a brown torso and the famous thick foot approach a burrow. A calculated swipe here, another there and the pair casually saunters away. For a few minutes, there's a deafening silence. Then the hide erupts in squeals of joy. The seemingly ordinary *jungli murgi* has won us over.

This piece is based on an original story commissioned and made possible by Roundglass Sustain, a not-for-profit that tells stories of India's natural world to create awareness and support conservation.

ABOUT THE AUTHOR

Radhika Raj has worked as a researcher, writer and editor for over a decade. She has written for some of India's leading publications, including the *Hindustan Times*, the *Indian Express* and the *National Geographic Traveller*. She is currently the deputy editor at Roundglass Sustain, where she writes and edits stories and scripts on India's biodiversity. She is also a South Asia Speaks (2024) fellow and is pursuing a project that chronicles the lost wildernesses of undivided Punjab.

Radhika has a master's in Sociology from the Delhi School of Economics and an MPhil from the Tata Institute of Social Sciences. In her previous life, she worked as an ethnographer, chronicling everyday life and violence along a city's margins.

Born and brought up in Mumbai, she currently lives in Delhi with her partner and two cats. When she is not chasing stories, she is dreaming of building a pottery studio in the hills.

8

Nong-in – The Bird that Tracks the Rain

ANITA MANI AND SHASHANK DALVI

From a bird that hardly anyone had laid eyes on to one that can be twitched on a short trip to northern Manipur (though whether you should attempt that now is another question) is a stunning change in status. But that's exactly what has happened to the Mrs Hume's Pheasant. Despite being the state bird of Manipur and Mizoram, and having captive breeding programmes at zoos, this beautiful pheasant remained unseen in the wild until an intrepid birdwatcher from Manipur, Harmenn Huidrom, changed the status quo. The key lay in the strategy used to find the bird. With the help of local hunters turned conservationists, Harmenn found the bird at night in its roost, instead of looking for it during the day. The story demonstrates how important locals – including hunters who are more attuned to the fauna of their forests – can be in locating a species. It also demonstrates that engaging with hunters is crucial to the conservation of a species.

In March 1944, a newly raised regiment of the Indian Army fought the first-ever battle against the Japanese on Indian soil. The Japanese were looking to get a firm foothold in India from the east, and Operation U-Go was launched with the goal of capturing Imphal and Kohima. At the village of Jessami on the Manipur–Nagaland border, the might of the Japanese 31st Division was met head-on by the soldiers of the first battalion of the Assam Regiment. The battle lasted from 28 March to 1 April, and the soldiers of the young regiment – drawn from the Ahom, Naga, Mizo, Kuki, Khasi, Garo, Lushai and Manipur tribes – fought relentlessly, with no thought to their own safety.[1] By holding off the Japanese troops for as long as they did, the soldiers of 1st Assam helped strengthen Allied defences at Kohima and influenced the outcome of the Battle of Kohima, described as 'Britain's greatest battle'[2]; the defeat suffered by the Japanese in the battle changed the course of World War II in Asia.

The village of Jessami is now back in the news and is the centre of many an Indian birdwatcher's quest to see one of Asia's rarest pheasants, the 'Nong-in', also known as Mrs Hume's Pheasant (*Syrmaticus humiae*). In Meitei, a language spoken in Manipur, 'Nong' means rain and 'in' means to follow, so literally, 'one who follows the track of rain'.[3] In other words, the name speaks of a species with keen weather sense.

Mrs Hume's Pheasant was first described by A.O. Hume in 1880, in the ninth volume of his journal *Stray Feathers*, where he paints a vivid picture of his search for the bird.

The quest was triggered when Hume noticed a plume of feathers (which denoted rank in the royal court) sported by the envoy of the Maharaja of Manipur: '... I immediately saw that it contained three or four long tail feathers of a Pheasant with which I was unacquainted.'[4] He was informed that the feathers belonged to the *Loe-nin-koi*, a Meitei[5] name for the Hume's Pheasant. But no one in Manipur had seen it, as the bird resided in 'pathless hill jungles' controlled by the Kamhows, a fierce tribe known to kill anyone they came across. Thanks to the interest taken by the Maharaja in the search, the full resources of the state were placed at Hume's disposal. But to no avail – the bird became more and more of a myth, and 'it began to be suggested that "there never was no such bird".'[6] But Hume persisted and finally, an effort was mounted in the southern reaches of the kingdom.

> 'At the south of the Manchar Lake we got together the most important officers of the country further south, and my Envoy made them understand that the bird had to be got. It was not distinctly said everyone would have their heads chopped off if we didn't get it, but a vague gloomy cloud of awful possible eventualities was discreetly left to veil the vista.'[7]

Finally, through a carrot-and-stick stratagem (more of the latter, though) applied on a bunch of Kamhows who had taken refuge in Manipur territory, one skin as well as one live male specimen were secured. Unfortunately, the latter, a tame bird that eventually became a camp pet, was killed in a fire accident at the camp. According to information that Hume obtained from the locals, 'these birds live in dense hill forests at elevations of from 2,500 feet to fully 5,000 feet. They prefer the neighbourhood of streams and are neither rare nor shy.'[8]

Hume observed that 'like its nearest ally [the Elliot's Pheasant], this bird is distinctly an intermediate link between the true Pheasants *(Phasianus)* and the Fowl Pheasants (*Gallophasis*), and I think that both may well be separated under the generic title *Callophasis*'.[9] The old Latin name for the bird was *Callophasis humiae*; the bird was then placed in the genus *Syrmaticus*, which contains four other pheasants that share the common traits of a long, white-banded tail and red facial skin. The common name pays homage to the ornithologist's wife, Mary Ann Grindall Hume.

Today's birdwatchers need to employ far less dire stratagems to get a glimpse of the bird, which is also the official state bird of Manipur as well as Mizoram. And that is mainly owing to the fieldwork done by Harmenn Huidrom – one of the state's finest birdwatchers – who has a knack for finding rare birds.

Harmenn's interest in wildlife was kindled early on by his father who would take him to Imphal's zoo and nearby 'wild' areas. The flashy pheasants on show at the zoo captured the young boy's imagination – Harmenn recalls being struck by the beauty of the Green Peafowl, an endangered species last seen in the wild in India in 1928, though sightings are now possible in Myanmar, Thailand, Vietnam and other parts of Southeast Asia.

Although his father passed away when he was five years old, the interest in wildlife stayed with the young man. Hunting pressure in the areas around Imphal had cleared nearby wild spaces of large species, but birds remained, though in low densities. The Imphal of Harmenn's childhood had a few dense stands of forests and hillocks, such as Langol Ching and Nongmai Ching (with 'Ching' meaning hill) – both are Reserved Forests – and that's where he would head, armed with his birding bible – Sálim Ali and Laeeq Futehally's *Common Indian Birds*.

It is intriguing that Harmenn's earliest bird memory is of a rare pheasant, and so, it seems fitting that he ended up discovering the most reliable sites till date for seeing Mrs Hume's Pheasant in India. But then, Harmenn has a knack for finding rare birds – in 2017 he found the Black-breasted Parrotbill, in what was possibly the second record for the state, and was also part of the group that conclusively identified India's first Grey-eyed Bulbul from the Moreh region of Manipur, close to the state's border with Myanmar.[10] The region also yielded the first Indian record for the Rufous-winged Buzzard[11] and a new subspecies of the Eurasian Jay. A lot of these birds were found in border areas that are also, strangely enough, ecological transition zones. The Manipur–Myanmar border is a biogeographic boundary as the mountains of Manipur transition into plains close to the border. Some of this area is also protected under the Yangoupokpi Lokchao Wildlife Sanctuary.

But it was the discovery of the locations for viewing the spectacular Mrs Hume's Pheasant, in 2021, that lit up Harmenn's name in the birder's hall of fame. It wasn't the first time he was seeing the bird though. In 2009, a year after he began birding seriously, he was in the interior parts of Manipur's Ukhrul District (to the northeast of Imphal), when he got a glimpse of the male pheasant's long, white-patterned tail in the daytime. Local hunters told him that the bird could be quite easily seen at night. Wary of venturing into the jungle in the dark when he wouldn't be able to photograph the bird, Harmenn demurred. Had the idea of spotlighting the bird with a torch occurred then, the rediscovery of the wild populations of Manipur's state bird may well have happened over a dozen years earlier.

Harmenn was not the first to rediscover the bird – which was photographed in India for the first time in 2016 at Pungro in Nagaland – but that sighting was not repeated, possibly because no one went there again; it however pointed to the fact that the bird was clearly surviving in Nagaland as well. There was also an earlier sighting by the ornithologist Ravi Sankaran in the first decade of this century, not very far from Pungro. Before him, a set of surveys between 1996 and 2004 by Anwaruddin Choudhury in Nagaland, Mizoram and Manipur threw up multiple locations where the bird could be traced through preserved specimens, encounters by locals and a handful of sightings by Choudhary and others.[12] It was evident that the bird was consistently present in these areas. But a reliable location for this rare pheasant continued to be elusive.

Harmenn's 2021 encounter was triggered by a press note issued by the village chairman of Razai Khullen, in Ukhrul North, who claimed that the area contained nearly 350 individuals of the rare Pheasant. Older reports suggest that the bird was regularly hunted for the pot in that area, though birds were also sourced by locals for the forest department, presumably for breeding, in response to a reward announced by the local government for a pair of wild Nong-in, in 2010.[13] In mid-July 2021, Harmenn and three of his friends – Oken Sanasam, Akoijam Priyojit Singh and Babie Shirin – visited Razai Khullen to verify the claim. On his request to the chairman to show him at least one (live) specimen, the latter marshalled over a dozen hunters, who were then sent out to comb the nearby forest-covered hills. The hunters fanned out at 1.30 p.m. and by 7.30 that evening, three birds were spotted. During the search, Harmenn and his companions were advised to wait at the chairman's house. 'They told us we may be too noisy!' he recalls. It was a three-and-a-half-hour trek to the spot where the birds were seen. 'When we got there, it was drizzling a bit and then one of the hunters came up to us [and] told us he could show us a male if we approached slowly. And then before us was a fully grown male; we were so excited that we forgot it was raining, and began changing lenses to capture a photo . . . After all, it was a lifetime experience!' There was a female around as well, but they didn't manage to get a photo of her.

The male of the Hume's Pheasant is a spectacular bird – about 3 ft long (large enough to make it a struggle for photographers in proximity to fit the bird into a single frame), with the bare red facial skin around its eyes contrasting with the metallic blue of its throat, neck and upper breast. Chestnut-coloured flanks, prominent white bars on its wings and black-and-chestnut bars on a silvery tail finish off a truly impressive bird. The subspecies found in India's northeastern hills and western Myanmar is the *Syrmaticus humiae*, while the *burmanicus* is found in the forests of south-western China, northern and eastern Myanmar and the northern tip of Thailand.

So, are these birds, found in less than a night by local village hunters of Razai Khullen, also snagged for the pot or their feathers? Luckily not. And for that one has to thank the state's captive breeding programme. Pheasants are easier to breed in captivity than most other birds. Keen to supply to the state zoo for this programme, the village committee has mandated a massive fine – around 20–30 thousand rupees – if a pheasant is killed.

Subsequently, Harmenn discovered the bird in other locations as well, including at Jessami, further north, the last village in Manipur before the border crossing into Nagaland. This discovery played out on lines similar to the first time. When word got out in the press about the discovery of Mrs Hume's Pheasant at Razai Khullen, people in Jessami started claiming that the bird was being seen in their area too. When requested to show evidence of the species, the village council ordered hunters to go look for the bird. It rained the first night and no birds were seen, but the next night's search delivered eight to nine individuals. But unlike in Razai Khullen, hunting was a problem at Jessami; one way of countering this was to project ecotourism as an alternative livelihood to hunting, an idea evangelized by Harmenn and catalyzed by others such as wildlife photographer Dhritiman Mukherjee. They guessed (correctly) that once word got out that the Mrs Hume's Pheasant could be seen in Jessami, it would trigger a stream of birders keen to tick off a once mythical species.

Both Jessami and Razai Khullen, where the bird was seen (and sighted repeatedly thereafter), are on the same range – the Naga Hills – but the birds are easier to see (read shorter walk) at Jessami, which is located close to a highway. This has made it a far more attractive proposition for birders hauling cameras and other equipment. Incidentally, the location of the 2016 sighting in Pungro, Nagaland, lies just over 100 km northeast (and possibly shorter as a bird flies) from Jessami on the same range.

The Mrs Hume's Pheasant is the state bird of Manipur, but it is hard to find cultural or religious associations for the species in local lore. A lone thread cited in the local media is the story that speaks of a young, much-in-love couple, who, when they were dying of hunger and thirst, transformed themselves into a spectacular pair of Nong-ins.[14] Other lore says that killing the Nong-in will transfer the sin of killing the couple to the killer.[15] Harmenn believes that the long tail feather was used to signify a soldier's rank in the local army/militia. The birds feature in few stories in local folklore; Harmenn however recalls being told, when he was child, that it was important to conserve it. While this thread runs strong in the communities living on the plains of Manipur, hunting in still prevalent in some hilly tracks, where the bird's size makes it a prize for the pot.

Today, ironically, local hunters turned bird guides are key to sightings; fantastically skilled at tracking inside the forest, even in deep darkness, hunters fan out to locate the pheasant once it gets dark, usually after 6 p.m., while visiting bird watchers wait at a convenient point. A team of four to five hunters will typically disperse to scan different areas. 'While you wait, you can see [...] flashes of torchlight [...] on the hills here and there, where these hunters are searching for the birds. Luckily, the phone network is good and the hunters communicate with each other and with us over calls and WhatsApp,' recalls Mumbai-based birder Amey K., who travelled to Manipur in the summer of 2023. But how do they know where to look? 'The hunters go to an area known for the bird in the evening and wait, listening for the sound of the flapping of the pheasant's wings. That is how they know where to go and look for it. Plus they look for the faeces, the droppings of the bird; they also know the call of the bird, in fact the different calls of the bird, by heart,' explains Amey. Once a location is ascertained, birders trek up to the point.

The local community at Jessami has been quick to embrace ecotourism and has even set up a Facebook page to promote the destination to birdwatchers. In January 2024, the Jessami Mrs Hume's Pheasant Community Reserve was inaugurated – this is part of an initiative by the local village council to conserve the species and its habitat. The reserve is located 15 km from Jessami, close to the Indo-Myanmar border. Six former hunters are now spotters for the Reserve.[16]

In addition to Jessami, Mrs Hume's Pheasant is now being seen at several locations in Manipur, including Khamasom and Shiroy Chingkha, thanks to Harmenn's efforts. When photos of the bird were circulated to hunters in habitats where they were likely to be found, some of them responded saying that they had indeed seen the bird in their area. Which brings one to the question – is the species truly rare in Manipur? The answer is – possibly not. Locals in Khamasom recall seeing the bird out in the open, close to their homes. But today, hunting pressure sends the bird into hiding at the first whiff of human presence, which is why chance encounters in the daytime are rare. But it can be tracked and located by those who know the forests, which indicates that the species is present in good numbers and is clearly breeding well. It is also highly probable that the bird may be present in other North East states, such as Mizoram, within a dense oak–pine mosaic. What makes this bird 'rare' is the fact that it can be reliably seen only at night, a fact that also makes population estimation difficult. Of course, you need to keep in mind that the bird is also rare in India due to the rarity of its habitat, which is intact conifer–oak forests.

The challenge confronting the conservation of the Mrs Hume's Pleasant is hunting; guns and catapults are the norm, and one would be hard-pressed to spot even a single primate in these hills, which have fallen prey to the empty forest syndrome. Amey recalls hearing shots go off on the hillside while staking out the pheasant. Given the sought-after status of the bird and the diversity of species in the area – Amey and his companions found a host of northeastern specialties, including Wren-Babblers, Hodgson's Frogmouth, Mountain Bamboo Partridges, Yellow-throated Laughingthrush, Striped Laughingthrush and White-browed Laughingthrush, en route to and around the Hume's Pheasant habitat – there is potential for bird tourism to provide livelihood alternatives to hunting. Like the hornbill nucleus economy in Latpanchar, a birding hotspot in North Bengal, the Mrs Hume's Pheasant has the potential to catalyse wealth creation in an area where subsistence living is the norm, where children in schools have seen common fruits like the watermelon only in books. A beginning in this direction has been made at Jessami.

Observed only at night, there is little information on the bird's behaviour in Indian forests. What is known, however, is that it favours tender oak shoots, a feeding preference it shares with tragopans, the Khalij Pheasant and the Red Junglefowl. Its habitat is conifer–oak forest, with grassy patches spread over steep slopes. For Amey, the landscape mirrored the Cheer Pheasant habitat in the western Himalayas, except that the forests of the Northeast (the Hume's Pheasant habitat) were denser. But unfortunately, 'the forests have been hunted [in] and burnt extensively,' points out Amey. 'They've not even allowed secondary forest to re-grow, and the problem is that the fire is not a controlled fire.' The result is bare slopes, mostly everywhere, with denser strands only as one goes towards Myanmar, especially where the Indian Army is present. Amey saw a pheasant in one such border forest, an hour east from Khamasom.

Like many other species in the family *Phasianidae*, the bird roosts on trees at night. Though it stays out of sight in the daytime, the bird is surprisingly tolerant of human presence at roosting time. Amey recalls how the hunters were rarely silent as they moved through the jungle, talking among themselves, speaking on the phone and playing music. Even when visibly watched through binoculars, the birds are not skittish; movement, as long as it is not sudden or fast, is also tolerated. 'But if you touch a tree, and even if the branch you are touching is not connected to the one the bird is sitting on, that little vibration on the trunk is enough to make the bird scoot off,' explains Amey.

Though it has so far been seen and photographed only at night, the Mrs Hume's Pheasant is not a nocturnal bird. It departs the roost just before dawn and is active during the day, but it has been spotted, so far, only during its night roost. In Jessami and nearby villages, hunting pressure has tapered off, and if this persists, we may finally get to admire the bird's splendid plumage by light of day.

The move away from the hunting of the Nong-in is not uniform across the region; Harmenn recalls waiting, along with a visiting birder from Mumbai and a local hunter, for the bird to show at another location on the same range. The latter was all set to take the target down once the sighting was done (the shortest lived lifer ever?) but luck, in the form of gender, prevailed – the male that the hunter was after failed to show. He, however, takes heart from the fact that villages like Razai Khullen and Jessami are taking the first steps towards conservation. The success with the Mrs Hume's Pheasant has triggered an interest in looking for other species that may also be found in these forests – typical South Assam species such as Striped, Yellow-throated, Moustached and Brown-capped Laughingthrushes, Spot-breasted Parrotbills, Crested Finchbill and Flavescent Bulbul. And that will be Harmenn's focus once peace returns to Manipur and he is able to take to the field again.

9

In Search of Vicky on Phawngpui

PUJA SHARMA AND ANDREW SPENCER

The interestingly named Mount Victoria Babax is a classic example of a rare bird that was re-found when someone made the effort of showing up in the right place and at the right elevation. Although the bird was seen and sound recorded previously, Andrew Spencer and Puja Sharma were the first people to not only find the bird in the blue mountains of Mizoram, but also to spotlight the species and its habitats for other birders to make the pilgrimage. This was a bird that was waiting to be discovered. What makes it special – it is not a flamboyant bird – is that it is restricted to just one hill range (Chin Hills) in the whole world; what adds importance to the Mizoram population is that Myanmar, where the eponymous Mount Victoria stands, has become inaccessible. A true case of geopolitics adding to the 'rare' quotient of a species.

There's something fascinating about the remote corners of the world. Off the beaten track locations have attracted a certain type of birder for as long as birding has existed. And India is blessed to have a plethora of such places to explore!

Unfortunately, the COVID-19 pandemic had put a hold on such explorations for Puja and I for nearly a couple of years; so it was with great anticipation that I was finally returning to India in late 2021, when the country was just starting to open up. 'Where should we go?' I asked. Sikkim? Arunachal? And so our discussions went on. Some Indian states that border China have strict rules and permits for visitors, and some of these remote locations were still not open for tourism due to the pandemic. We were beginning to become resigned to our fate of being stuck without a proper birding plan – be it a plan A or plan B.

But all that changed during a long phone call with our dear friend Dolly Laishram, when we lamented to her how none of our plans seemed to be materializing. The previous winter, Puja and Dolly had spent nearly a month birding and exploring some of the valleys around Imphal, lodging in Dolly's ancestral village, Nambol, at her family-run birding resort, Masha Homestay. 'Masha' in the local Meitei language means feather or leaf. And true to its name, it was an idyllic, charming countryside retreat, abounding with fruiting trees, leafy mustard greens, bamboo clumps, a deluge of birds and a soundscape punctuated with the echoes of Striated Grassbird song from all directions. The house was surrounded by paddy fields, boggy swamps, wetlands, waterfalls, distant valleys and hilly vistas. All this meant a myriad of 'birdy' habitats to choose from, and a slew of hopes and dreams of ordinary and extraordinary Manipuri birds.

Ever since Puja's trip and her endless stories of Manipuri birds, I had wanted to experience its unique birdlife for myself. To see those species lucky enough to have *manipurensis* as their epithet. And maybe even, if one could dream, to stumble on a Manipur Bush-Quail. So, when Dolly told us to simply book tickets for the three of us to Imphal, we did as we were told. One thing led to another, and one fine day Puja exclaimed, 'How about we touchdown in Manipur first, *and then* Mizoram?' Immediately, the thought of those remote corners and their rarely seen birds began to cast their spell on me.

The first part of our trip – with the three of us donning our masks and face shields on the plane – to Manipur is worthy of a chapter all on its own. Puja and Dolly, along with their friend Harmenn Huidrom, had recently documented the first records of the Chestnut-crowned Bush Warbler for Manipur[1]; and earlier that month, Harmenn had stumbled upon a mysterious *Iole* bulbul[2] near Moreh, on the border with Myanmar. Looking at range maps and the relative paucity of eBird records, it was clear that there was a deep well of additional potential in visiting the area. So, the probability of finding ordinary birds doing extraordinary things was enormous. And the chance of finding extraordinary birds and contributing information to the avifaunal repository of this unique landscape was prodigious.

With these objectives on our minds, we set out first to Kwatha village, which was barely a hop, skip and jump away from the Burmese border. There we found that the mystery *Iole* bulbul best matched the Grey-eyed Bulbul, while a fortuitously photographed mystery buzzard later proved to be the Rufous-winged Buzzard – both country-first records for India and South Asia.[3,4] But the excitement didn't end there! A distant jay, when examined more closely, was not the expected 'Plain-crowned' subspecies of Eurasian Jay, but rather showed features of the 'White-faced' group, hitherto undocumented in the country.[5] Later, research showed that these birds (the ones that we had found), in addition to the others nearby in Myanmar, might represent a hybrid zone that is, as of now, poorly documented.

A couple of Collared Falconets along the road meant the firsts kept on coming – a first for Manipur in this case.[6] And finally, we also stumbled upon a mystery. Several nuthatches in the area were immediately noticed as looking 'off' and not quite like the widespread Chestnut-bellied Nuthatch. Doing our best to document them, we took a detailed series of photos that showed they could very likely represent a range extension of the Burmese Nuthatch, another potential first for India.[7] This last sighting also goes to show that not every mystery has a satisfying ending. We don't yet have the definitive answer. But that's okay! Mysteries like that are one of the things that add some spice and excitement to birding.

With our Manipur explorations coming to a close, we felt as though we had been anointed by the Manipuri 'bird gods', so to speak. But it was Mizoram where we would be devoting the lion's share of our time. With a location tucked away closer to Vietnam than to Delhi, the hills of Mizoram exude a mystique that is hard to match. Here lay the heady mix of potentially rediscovering a missing species from the national bird list and the remarkable opportunity of mining the treasure chest of some of India's rarest birds. One species above any other was emblematic of this – the Mount Victoria Babax (*Pterorhinus woodi*).

This species, essentially a fancy laughingthrush, was considered conspecific with the Chinese Babax for many years. Its global range is tiny and stretches from Mount Victoria in Myanmar – over a distance of barely 100 km – and just about tiptoes into India. Similar in size to most other laughingthrushes, it is distinct in its strongly streaked plumage, dark cap and whisker and piercing pale eyes. And, like most laughingthrushes, it is most easily found by voice – a sound we would be hearing over and over again soon enough as we searched high and low for this mysterious bird!

To understand the draw of the babax, one also needs to know a little bit of the history surrounding the occurrence of the species in India. Found only on one mountain in one of the most remote corners of the country, nothing about the Mount Victoria Babax could be called 'easy'. It was first discovered on Phawngpui way back in 1953, when an expedition led by Walter Koelz and Thakur Rup

Chand visited the area and collected seven specimens over the course of three weeks, between 29 March and 18 April 1953.[8]

What followed is shrouded in some mystery. The species was certainly not seen again in India until a 1997 study on the mountain by Dipanker Ghose, but even then it was relegated to a checklist of a single study published in an obscure journal, without any further details.[9,10] It was seen again and sound recorded in March 2016 by Indian Forest Service officer Pratap Singh, but again without any publicity or fanfare, or anyone else even being aware of the record.[11] In essence, prior to early 2022, the species rightly deserved its place amongst the 'hardest species to find' in India, as well as being one of the longest 'lost species' on the national list of birds.[12] With these tantalizing facts in mind, it was little wonder that the babax was on the top of our 'target list' for the trip.

Flying into Aizawl, we were already impressed by the sheer scale of the forest covering the knife-edged ridges north of the city, salivating at the thought of all the birds that must be there! It didn't take long after our arrival to see that our initial aerial impression was well borne out. Forest-lined roads filled with birdsong, and that was just on the drive from the airport to our first hotel, the government-run tourist lodge at Lengpui.

After we were picked up by Joe R.Z. Thanga, our indomitable and indispensable ground agent, local guide, driver and, also, an all-round great company, our first major port of call in Mizoram was the foothill town of Sailam, a few hours' drive from Aizawl. It became increasingly clear as we drove that our first impressions from the plane were a 'Mizo theme' – the landscape was forest-heavy, and the forests were bird rich. But what also quickly became apparent was that the local communities were an integral part of that. One person's remote corner of the world is another person's homeland, and it was clear that those lucky enough to call Mizoram home have been excellent stewards. The community-run ecotourism infrastructure of Sailam was an excellent example of this. We were warmly received by the residents of the village, both young and old, and made to feel at home in the shelter of the aptly named 'Birders' Cottage'. This was part of an ecotourism project undertaken by the locally run Sailam Ecological Conservation Society, intended to promote community livelihoods through conservation and ecotourism initiatives. Over the next couple of days, with the excellent help of our local guide Ruatafela Mafela, we were treated to roads and trails through well-protected forests and birdlife undisturbed by hunting. It was greatly heartening for us to experience how much everyone there cared about both their environment and wanting to share it with others.

From Sailam, we drove to our next destination – Sangau, a tiny hamlet at the foot of the majestic Phawngpui. After a long day of difficult driving, manoeuvring landslides and roadblocks, we eventually reached a barricaded sign of 'No entry to Sangau' late in the evening. Little did we know that the local village council had ordered the village residents to self-isolate and had barricaded the entire village from outsiders in the wake of increasing COVID-19 cases being reported in the country. Thankfully, Joe was able to successfully arrange a meeting with the village council and, after proving that we were healthy and safe to be around, the village head allowed us to proceed with our plans.

Even by the standards of a country as amazing as India, Phawngpui is a special place. That became clear as soon as we crested the final ridge below Far Pak and first laid eyes on the expansive open valley below us, bracketed on three sides by stunning forests and on the fourth by a sheer cliff that took our breath away. The isolation was palpable – as far as we could tell, not a single other soul shared the mountain with us. It was hard not to whoop with joy at the overwhelming beauty of it all, but at the same time it felt wrong to utter any sound that would break the serenity. And as icing on the cake, somewhere out there was that babax – the bird we had made all this effort to try to find! But before we could rush out and begin our babax quest, we had to get ourselves situated. Luckily, despite its remoteness and isolation, Far Pak – a grassy meadow nearly half-way up the summit of Phawngpui – comes with a forest rest house. And we were fortunate to have a roof over our heads and a place to set up for our stay.

Our hope of quickly finding the babax was soon squashed during our first day itself on Phawngpui. But any disappointment we may have had was very quickly forgotten, for it was immediately apparent from our first minutes of dawn on the mountain that birds were everywhere! Even though the dawn was cold, with frost coating the ground in some of the depressions, there were birds singing from all corners of the valley. Being sound recordists, we were immediately entranced, not knowing whether to record the Striped Laughingthrushes in one bush or the Buff-throated Warblers calling from the tall grass near the rest house, or one of the many Crested Finchbills high in the pines outside the front door of the rest house.

If the morning cacophony wasn't enough to demonstrate how wonderful a place we had found ourselves in, a trickle, turning into a stream, of thrushes flying overhead drove the point home. Now why would a flock of thrushes do that, you might wonder? Because the very first bird we put binoculars to was not the expected Eyebrowed Thrush, but rather the much rarer Grey-sided Thrush! And so was the next, and the next, and the one after that. And while we did also find some of the more common Eyebrowed Thrushes, all in all we counted over forty Grey-sided Thrushes, an astounding total for this very local and uncommon species.

The highlights kept coming one after the other. Stripe-breasted Woodpecker? Check. Slender-billed Scimitar-Babbler? Check again. Spot-breasted Laughingthrush, Hume's Treecreeper, Chin Hills Wren-Babbler, all quickly found, seen beautifully and well recorded, rounded out the first hour of daylight.

Not finding the babax around Far Pak, we decided to tackle the peak itself. Not only would this provide the longest expanse of potential habitat to search, we had dreams of maybe, just maybe, finding something even rarer. Could some other Burmese endemic be hiding on Phawngpui, waiting to be found by the intrepid birder? We hoped to find out!

Led by the intrepid Chhankunga Zathang, our local guide who knew this mountain and her forests like the back of his hand, our march to the peak brought us through kilometre after kilometre of beautiful, untouched habitat. Habitat that looked perfect for a babax, but alas did not provide. However, it was part way up the mountain where we found what has since proven to perhaps be our rarest find – in the midst of a huge, impenetrable patch of bamboo, a fast-moving, noisy flock of 'Buff-breasted' Black-throated Parrotbills absolutely delighted us with their playful antics as they cavorted around us! This distinctive subspecies has a very similar range to that of the babax and has been seen by very few birders in India.

Finally, breaking through the last bit of dense forest and summiting Phawngpui was almost as transcendent as finding the parrotbills. It was hot, sunny and nearly birdless, with just a few Himalayan Griffons lazily flying by overhead keeping us company, but man oh man was the view to die for! Forested ridges as far as the eye could see, hill tops of three different countries (India, Bangladesh and Myanmar), and the feeling of being on top of the world at the summit of the mountain made the previous kilometres of tough hiking just melt away.

All good things come to an end, however, and it was back down, towards Far Pak, that we spent the rest of the day. As always, we searched high and low for the babax. And as always, we came up empty. Thankfully, all the other birds we had become familiar with kept us company, and some of our best sound recordings of the entire trip rounded out the afternoon.

Our final full day on Phawngpui dawned as those before – clear, cold and filled with birdsong. Fire-tailed Sunbirds were abundant at the edges of the meadow, the morning stream of Grey-sided Thrushes continued to amaze us while Aberrant Bush Warblers chattered angrily at our feet, and some more excellent recordings of the various laughingthrushes, scimitar-babblers and assorted mixed-species flocking birds filled our memory cards. But the babax? Not a sniff or a sign. To be honest, I was beginning to lose hope of finding the bird at all. Had they disappeared from India since the days of Koelz, we wondered? Like everyone else, we were unaware of the more recent sightings, so the 'disappearance' did seem to be a reasonable explanation as to why we were coming up empty.

But it's hard to be too down when you're out watching birds. Doubly so when you're encapsulated in the grandeur of mountains and forests, and all of the splendid birds that live there. And even though the day came and went babax-less, Phawngpui still rates amongst the favourite places we had travelled together. Our last night on the mountain was filled not with the sadness of thinking that we had missed our main target, but instead with an unwillingness – for having to leave so soon.

So it was that another cold, clear morning dawned, and we were packing to leave. We had just put away the last of the breakfast supplies and our backpacks were latched closed. Puja, as usual, was glued to her binoculars, trying to catch a fleeting glimpse of a covey of Mountain Bamboo-Partridges scurrying away. So I walked a few meters away from the rest house to take in Far Pak one last time, and listen to what the dawn chorus had to offer. Striped Laughingthrush, check. Buff-throated Warbler, check. And then, right at the very edge of my hearing – wait, that sounds kind of like the babax! But no, surely not. It would be like the plot of a bad movie to get the target at the last minute of the last hour. Surely my mind was playing tricks on me?

The more I listened, the more certain I became. Sprinting back, I yelled out for Puja to come and listen. But then the song stopped, and again doubt crept in. But nothing ventured, nothing gained. The two of us threw on our backpacks, said goodbye to the bemused Joe and, telling him we'd see him down the mountain, hoofed it towards what we heard. And then, there it was again, closer and louder, and we were nearly certain. A babax, singing naturally, right by the edge of the precipice of Far Pak!

After making sure to get a witness recording, we snuck closer and closer, listening to the tantalizing song of the babax, finally heard after all our efforts. But we still could not lay our eyes on it. In the end, we walked nearly 800 m from the rest house before finally spotting our quarry, sitting calmly in a large bush, as if to say, 'Where have you guys been? I've been calling out for you all morning.'

It's hard to describe the feeling of finally finding success after giving up hope, of being rewarded for coming so far and after a long, hard search. Our elation was tempered as we spent the next half hour documenting the babax, vacillating between wanting to yell out in joy and trying to keep as still and quiet as possible while recording, photographing and filming the bird. In the end, though, the bird finally decided that it had had enough of us and descended the side of the cliff face where we couldn't follow, and we could now step back and finally absorb what had just happened. That surreal moment of knowing we had found what we came for, made all that much sweeter for having had to work for it. Joe and Chhankunga had already left Far Pak and started down the mountain, so it was just the two of us, all alone on Phawngpui, surrounded by grasslands and forests and that stunning vista, and yes, even a babax!

Eventually, though, we had to come back down to earth. An obligatory quick burst of WhatsApp messages, with some screenshots of our photos to some of our friends, and then we followed Joe down Phawngpui, the rest of our time in Mizoram beckoning. Though to call the trip down a 'hike' would do it a disservice — it felt more like we were floating, buoyed by our success and reliving the moment over and over.

One major location remained for our time in Mizoram — Murlen National Park. While we had no particular major targets here as we did on Phawngpui, Murlen is no less amazing a place. Keeping with the theme of Mizoram, forest abounds and birds are everywhere, along both the roads and trails. Some rare species are even more common here, among them the Stripe-breasted Woodpecker and Chin Hills Wren-Babbler. It was a fitting end to our journey to this remote corner, one we hope to return to someday soon!

In the days and weeks to follow, the story of the babax came into focus.[13–15] As it turned out, ours was not the first sighting since 1953 when Koelz and Thakur had discovered the species and collected their specimens at Phawngpui, as we had initially supposed. Others had recorded the bird a mere few years before, and again some years before that. But since knowledge of these sightings was known only to a few, the true impact of sighting such a special bird that is not found anywhere else in the country was not fully felt. In the end, rather than saying that the babax was refound, we were happy to take our place as its newly appointed press agents instead, putting a spotlight on the positive impacts birding can have in these magical, remote corners of the world.

Now a regular stop for many, not a spring has gone by without Facebook posts filled with photos of birding groups visiting Mizoram. Compared to our four-day search on the meadows and forests of Phawngpui, and the last-minute save, the babax has been 'figured out' and typically found quickly and multiple individuals are often seen. In addition, birders have continued to find more and more reasons to visit the area, beyond the single headlining species. As an even increasing number of people have been able to share in the wonder that coming to amazing places, such as Phawngpui and Mizoram, can bring, it is our hope that this trend continues to grow, and becomes a testament to how birds and birding can help showcase what an incredible world we all live in. In today's age of an ever-shrinking world, with nature constantly under threat, that can only be a good thing.

ABOUT THE AUTHORS

Puja Sharma is an ardent bird sound recordist from India. She saw her first Siberian Cranes at the Bharatpur Bird Sanctuary when she was nine years old, and has been fascinated by nature, especially birds, since she was a child. While living at her grandparents' house in Mussoorie, in the western Himalayan foothills, she found her profound love for bird sounds and has been enthralled by bird vocalizations ever since. Nothing gives her more joy than birding and hiking the mighty Himalayas, and pointing her microphone towards a bird until it has been heard (and recorded!) to speak for itself. A natural history enthusiast, she has a keen interest in studying historical bird literature and specimens and mapping bird records of the Indian subcontinent.

Andrew Spencer has been pursuing a lifelong love of birds ever since his grandmother showed him a Wood Duck when he was four years old. After spending his formative years exploring the United States, his first trip to South America converted him into an ardent sound recordist as well as a travel addict! Following a six-year stint in the bird guiding industry, he switched gears and began working at the Macaulay Library at the Cornell Lab of Ornithology. It was this job that brought him to India for the first time in 2019, and not a day goes by when he isn't looking forward to his next trip to the subcontinent.

10

Lusting for a *Locustella*

JAMES EATON

The Long-billed Bush Warbler's story is one that finds an echo across the world in an era of mass extinctions. Once extremely common, human domination of the landscape – even in the high elevations of the Himalayas – has all but made it impossible for the species to survive. Its hold on its habitat is so precarious that the species' remaining population may get decimated in a single lifetime (ours)! As James Eaton, the birder who explored the scenically spectacular Gilgit–Baltistan area of Pakistan-occupied Kashmir to find the bird, writes, the weakest link in the bird's survival lifejacket is clearly visible but the question is – will we be able to do something about it? For Indian birders, who may not be able to make it to Gilgit-Baltistan, the target may not be out of reach, for the bird may be hiding somewhere in the Kashmir Himalaya for another intrepid birder to find.

I am one of those classic, stereotypical, manic birders, suffering from obsessive, compulsive and nervous tendencies. The kind wanting to be in the field all the time. Every time I attempt to relax, take it easy, try a non-birding holiday, it simply doesn't – can't – come to fruition. From an early age, I've wanted to go and explore, find out things for myself, ever since my grandmother showed me the illustrations in Benson's *The Observer's Book of Birds*. This has well and truly poured over into my birding life – whether it be employment (I run a bird tour company, chasing birds), holidays (chasing birds) or reading (on early-age exploration by our contemporaries). Do I have a hobby outside of birdwatching? Sure, photographing birds!

One irritating trait I have is the tendency to get hooked on specific species, with a *need* to see it, no matter what it takes; I know somehow, someway, that 'moment' will happen – most of the time – as it has so often in the past. My compulsive nature has led me to pursue numerous lost species – birds that haven't been seen for prolonged periods of time – as that feeling of laying your binoculars on the prize is an addiction greater than what any narcotic could give you. What led me to realize this for the first time was when, way back in 2005, three of us decided to camp in an old, abandoned national park building on the shores of Ranamese, a crater lake high up in a cloud forest on a remote island in Indonesia. We hitched a ride there and decided to stay a few nights until we could find the vocally unknown, and overall little-known, Flores Scops Owl. To cut a long story short, despite wandering around, we drew a blank, though (unknowingly at the time), we took the first-ever field photo of the huge, arboreal Flores Giant Rat, a rodent known only from a tiny number of observations, and the only remaining extant species from the genus. Then we heard a Red-legged Crake in the middle of the night – at the time this was a new bird for me, so in attempting to find it, I shone my torch towards where the noise was emanating from – which was oddly from the canopy – only to illuminate a tiny, rufous owl, with big beady yellow eyes staring down at me! There it was, the Flores Scops Owl. And with it came that rush of excitement that gave me the kick that I will never need to seek from heroin.

This put me on a mission to find lost and new species. Living in Malaysia, Indonesia was the natural choice for such a search, given that this vast archipelago still had large tracts of seldom-visited forests and islands. The proximity also meant that forays to Indonesia could be both easy and frequent. What made it the perfect choice was the fact that, throughout Indonesia, were a host of species not seen in this century (many of which would turn out to be low-hanging fruit), while there was also plenty of room to discover taxa unknown to science.

I ended up with some remarkable finds in Indonesia, such as stumbling upon a hitherto unknown and undescribed parrotfinch, completely by chance while taking a 'break' in 2012. This was to be the first of several new and rediscovered species I was fortunate enough to find – including the Critically Endangered Boano Monarch, which was rediscovered on a tiny island – and culminated in the discovery of multiple new birds, then unknown to science, in an isolated mountain range in Indonesian Borneo. These highs are also matched by lows; the latter discoveries in Borneo happened while 'failing' to rediscover a long-lost babbler – the Black-browed Babbler – unseen for 170 years, only to find out, a couple years later, that we were looking on the wrong side of the mountain range! Or another owl chase, once again on a tiny island in Indonesia, where another bird, Siau Scops Owl, lost for more than 140 years is still waiting for someone to discover it; that won't be me as my protracted stay on the island in search of it resulted in zilch. It's the hope that kills you, or that which keeps you trying!

Some ten years ago, my OCD gaze fell upon a lost, little-known, little brown job (LBJ) – the Long-billed Bush Warbler (*Locustella major*) also known as the Long-billed Grasshopper Warbler.

Firstly, a bit of background on what exactly this 'LBJ' is, that has literally crept under the radar all these years. A member of the family *Locustellidae*, the genus *Locustella* comprises about 23 species, depending on your taxonomy, which are found breeding through Eurasia, and South to Southeast Asia, with a single outlier in Africa. All are largely migratory, barring the tropical species. Like the rest of the genus, the Long-billed Bush Warbler is an all-brown small passerine, with long undertail coverts, a broad, rounded tail and a chunky body. The plumage is variable; some have little black markings on the throat, some do not. Some have an all-black bill, some quite pale, and that's about as exciting as it gets to look at. The species is considered a migrant, though with no records away from the breeding grounds, we can only presume it winters in the lowlands immediately south of where it breeds, as this is what similar species do.

Historical literature on the species paints a very different picture to today's landscape. It is known only from a contentious border area, the Line of Control (LoC), in both India and Pakistan, and also just across the border into China (a single specimen). However, 99 per cent of the records are from long before the LoC even existed. From the 1880s to the 1930s, many ornithological figureheads, including John Biddulph, James Davidson, Bertram Osmaston and Edward Charles Stuart Baker, used statements like 'very abundant', 'very common' and 'calling all day long' to describe the species status.[1] Describing the species, Davidson wrote: 'Very abundant among the long grass arid weeds fringing the forests . . . We did not see it till the 8th June, when in the evening we heard its perpetual tic-tic-tic in the dusk. By the 10th it was very common and calling all day'.[2] So why should such a bird pique my interest?

Well, it's one of a number of bird species in the region on which all recent literature is woefully ill-informed. Given the 100-year-old information, when I started looking into it, I realized that there hadn't been any certain records of the bird since the 1930s! Adding to its allure, a photograph appeared in print, in Kennerley and Pearson's Reed and Bush Warblers monograph,[3] no less, from Himachal Pradesh, which was actually some sort of *Acrocephalus* warbler; so, as it turned out, I couldn't find any records of the Long-billed Bush Warbler for the previous eighty years in India, and it had never been photographed! Across the border, in the Gilgit-Baltistan area of northern Pakistan, the last record was in 1996, in the Naltar Valley, a highly remote area which, when I first started gaining an interest in this bird, didn't appear to be a safe area to visit as a foreigner.

This species was clearly ripe for a rediscovery.

My close friend, Shashank Dalvi and I discussed the Long-billed Bush Warbler (and many other long-lost birds of the region) and how I was going to find it, back in around 2012. Well, rather annoyingly, in 2015, he went on ahead and ended up obtaining fleeting views of a couple of individuals of the species on a fine, late summer's day near Kargil, right in the very valley where it was 'very abundant' eighty-five years ago. Was I happy for him? Of course not! This was my bird, and I wanted to find it! However, as Shashank had not obtained any photos or sound recordings, his sighting would likely be relegated somewhat in today's age of digital birding – much like I am doing now! Armed with his coordinates, and after looking intently at satellite images on Google Earth (my go-to for remoter birding areas), I visited the exact same area the following July, in 2016. However, I failed, despite trying in the same spot and the immediate vicinity; the heavily grazed landscape looked noticeably suboptimal to what I had been expecting.

In the following years, I returned in the summer months to the Vale of Kashmir, always with an eye out for this bird but, even till today, even as I write this, there are no further records from Indian Kashmir, despite numerous eyes and ears having scoured the landscape.

As Pakistan has had a rather turbulent recent past, I had been reluctant to visit. However, the pull could only be resisted for so long and, in June 2022, I made the decision to go in search of the bird there. The locality to search was obvious – the Naltar Valley, along the famous Karakorum Highway, deep in Gilgit-Baltistan, renowned for its breathtaking landscapes.

The Naltar Valley has a strong history with the bird, going all the way back to Biddulph's sighting of the bird, then again in 1986 and finally in 1996, when a friend, Dave Farrow, found four singing birds here. Since then, nada – though has anyone actually tried? Usually, it's a 'no' – birders generally are a bunch that enjoy treading the well-trodden path, rarely straying to explore new areas, leaving the less-explored areas open for those with a sense of adventure.

I even managed to convince my reluctant partner, Nancy, that we would have a lovely holiday, mixed with a bit of birding and wildlife watching ... Little did she know that for one, the trip was entirely centred around rediscovering the Long-billed Bush Warbler (well, she would have had an inkling, I suspect) and for another, that northern Pakistan had had a turbulent recent history, particularly in one area, very close to the LoC, which we would be visiting – of course, the police escort that was insisted upon would be the giveaway here!

Naltar Valley is a verdant, fertile valley high above the Hunza River; within lies a bustling hamlet, filled with domestic tourists who come here in the winter to ski, and in the summer for the scenery. It is a classic high-altitude landscape – snow-capped mountains tower over the periphery of the valley, grey rocks and crags eventually give way to a smattering of juniper, then to shallow terraces under cultivation, while emerald-green lakes and tall conifers are found further up the valley, away from habitation.

To reach here from the nearest major city, Islamabad, is quite the journey, and one we had no appreciation of in the slightest. A direct drive takes a couple of days; the first quarter of the journey is on a six-lane superhighway through the Himalayan foothills, before it suddenly switches to the classic windy, Himalayan roads we know so well from India, which are slow and traffic-laden. However, there is one huge difference between here and India – the lack of *BEEP BEEP BEEP*! As a foreigner, I'm unable to drive in India, which is perfectly fine by me, though I've always found the lack of control in one's destiny quite disconcerting. Tragically, we all seem to know someone who has lost their life on one of these roads. Therefore, I was apprehensive about being able to drive as a foreigner in Pakistan. I had totally and incorrectly assumed that the driving would be similarly chaotic and boisterous. How wrong I was - it was pleasant, mild-mannered, patient and, above all else, quiet!

The road to Naltar Valley eventually winds its way up to the 4,173 m Babusar Pass, which only opens in early June due to snowmelt. From here, the change in the landscape is dramatic, as the lush, conifer-clad Himalayas give way to the arid, desolate Karakorum. Deep, rocky gorges with peaks penetrating 6,000 m literally surround you. The famous KKH (Karakorum Highway) continues to lead you north, into Gilgit-Baltistan, along the Gilgit River, where irrigation has created fertile communes. Poplar and sea buckthorn surround terraced cultivation and hamlets. Every stop we made there would trigger a clamour for a selfie – we both felt truly welcomed in 'Selfistan'. With an encouraging wave to onlookers, we would end up with scores of locals in the selfies; Nancy also encouraged the shy ladies to take a selfie with her, leading to smiles on beaming faces afterwards, all under the glorious sun and towering peaks.

Once at the famous Silk Road trading town of Gilgit, you continue north once more along the more minor Hunza River, before taking a final turn, to the west, up into the Naltar Valley. The single road that winds its way up the valley was under construction, and it took us several hours to reach, as our Landcruiser revved its way up carefully. The trip was already quite the adventure, and hundreds of workers would stop and watch as vehicles tried and failed to make their way up; it was with a big roar that the Landcruiser made it up – nothing was going to stop us. We arrived at our lovely little guesthouse, the Jumaira Resort, in the afternoon, and informed the owner of our avian intentions, much to his bemusement. I don't think he had heard of a birdwatcher before. We did discuss the trophy hunting in the area however, and an ex-hunter-cum-guide visited us, pointing out where we should scan for the Himalayan Ibex and Markhor, two ungulates we were keen to see.

So, armed with a rough idea of where the Long-billed Bush Warbler was seen 26 years ago and a bunch of coordinates for likely-looking habitat from Google Earth, off we went the following morning. What is well known is the bird's liking for sea buckthorn and rough edges in and around terraced cultivation. We walked initially along the sea buckthorn–dominated riverside, though the noise from the turbulent river meant we soon veered up along the paths that cut through the terracing. One instantly noticeable feature was that the steeped terracing was now largely stone-built, no longer with mud and overgrown vegetation. Surely, this would have a big impact on the availability of habitat? As we walked several kilometres, we wandered through habitat I assumed would be perfect, but there was still no sign of the bird. Still, the landscape was breathtaking; snow-capped peaks high above us made way – through a transition zone of rocky slopes – for a lush, green valley, full of smiles and offers of tea. After drawing a blank in the morning, we headed up the rutted track to try a higher elevation. The habitat changed a little – to a heavily grazed landscape in which terrestrial birds would have a hard time. Himalayan Rubythroat, Mountain Chiffchaff and even Blyth's Rosefinch – the latter a new bird for me – were about, while we enjoyed hearing the frequent song of Brooks's Leaf Warblers, a rarely encountered bird on its breeding grounds up here and in neighbouring Afghanistan.

Realizing the habitat wasn't altering much, we went in search of something lusher. By early evening, there was one final waypoint to check from my virtual birding on Google Earth – a secluded area of cultivated terracing. It took a while to reach there, the narrow road, compounded by yet more construction making it difficult to reach. However, once we did, the difference was noticeable; we could see thick areas of *Rumex* (a type of herb) separating the small, ploughed areas planted with potato seedlings. Almost instantly, a response, a rhythmic 'tic ... tic ... tic ...' was barely audible in the distance. Elation! I impulsively ran towards the source of the sound, which must have been some 200 m away. Veering off to the left, I heard shouting behind me, it was Nancy, pointing to the right. Frustratingly, just a couple years before, I had lost much of the hearing in my right ear, with total hearing loss above 3 khz; since the song of the *Locustella* was at 4–8 khz, I was following my left ear, as these higher-pitched vocalizations are always 45 degrees to the left, in my mind! I would have got there eventually, but I would have circled round it a couple times, no doubt! We headed the right way and, very quietly, tip-toed over, delicately placing the speaker in a tiny gap in the *Rumex* where the bird *might* appear. I informed Nancy just how secretive almost all *Locustella* species are, and that we'd be lucky to get a flash of brown dashing across the tiny gap. We sat in silence, used some playback, and a bird flew straight out of the vegetation, sat on top of a large dock and sang right out in the open to its heart's content! Some 2,000 photos later, our satisfaction was complete – it was well and truly nailed!

Returning the following morning, I wanted to survey the area to see how many other birds there were, as I had heard a second bird in the distance the previous day. On arrival, one bird was actually feeding in a ploughed field, while two others were singing. All three varied quite noticeably in plumage, along with differences in the colour of the bill and throat streaking. Video, photos and sound recordings were all obtained; it was one of those red-letter moments to cherish – this is what birding is all about.

What have we learnt since? Having finished in the field and gained more of an interest and insight into the plight of the species, I wrote about its history and my sighting in an article published in *Birding Asia*, a journal published by the Oriental Bird Club, later the same year. During my research, it turned out that I wasn't actually the first to obtain a photo in the field – I unearthed a previously unpublished photograph by Charles Williams, taken from the Indian side of the border in 1977.[4] I contacted Charles, and he managed to digitalize his photograph, thus adding an important piece to the jigsaw. Following its publication, another record appeared the following year, 2023, in Gilgit-Baltistan again, but this time in an area closer to the LoC, by the village of Askole, close to where one would begin their trek towards K2, where one or two birds were heard and photographed. I also returned to the Naltar Valley in 2023, finding just two birds this time at the same spot, while a third bird was found a week later in the sea buckthorn bushes further up the valley by other visiting birders. This goes to show the importance of publishing your findings, as it encourages others to go out and look, and opens our eyes to the true status of little-known birds.

The species is classified as Near Threatened on the IUCN Red List of Threatened Species at the time of writing (with a proposed uplisting to Endangered base on research)[5] – meaning it's not a species we should be overly worried about in the short term, at least. However, it has become clear that we should worry about it. What can be done? Clearly, grazing by domestic animals is a huge issue throughout the foothills of much of the Himalayas and Karakorum. How can this be mitigated? This was not the only lost species of the Himalayas – the Himalayan Quail, a bird not encountered since 1876, probably met its demise due to habitat degradation for this very reason. Though grazing is the go-to for the blame game with habitat degradation, we also noticed another potential issue for the warbler – the *Rumex* our birds were found in had been sprayed with a weedkiller to kill the very plant the warblers are clearly dependant on; over the summer the plant would wither and die, right when the birds would be breeding. A survey and habitat awareness campaign has now begun in the Naltar Valley and the wider area, conducted by Pakistani conservationists and hopefully as their surveys spread, the goal will be to find more Long-billed Bush Warblers in Gilgit-Baltistan and raise awareness about the bird, along with ways to mitigate habitat degradation in the hope that we can turn the tide before the species is lost for good.

Now please, someone find that Himalayan Quail!

ABOUT THE AUTHOR

Born in Derbyshire, a land-locked county in the UK, it didn't take James long to spread his wings. Brought into the birding world by his grandmother, his interest was furthered by the help and willingness of his parents and lifelong friends met through the Young Ornithologists Club. The progress continued during his first birding trip to India at 17, which involved skipping college to indulge himself in the rich birdlife of Goa.

Residing in Malaysia since 2005, James has built an intimate knowledge of the identification, vocalizations and behaviour of birds throughout Asia. This has resulted in the discovery of several undescribed species to science, all in Indonesia, as well as numerous rediscoveries.

The culmination of this field time was the publication of over 100 manuscripts and articles over the past 20 years and lead-authorship of the *Birds of the Indonesian Archipelago*. James has a PhD in social sciences, having studied taxonomically cryptic species and their conservation status.

11

Twitching Tales

ATUL JAIN

For those who have been birding in India since the 2000s, Atul Jain is the original twitcher. The online Oxford dictionary defines 'twitcher' as an informal term for 'a birdwatcher whose main aim is to collect sightings of rare birds'. The key word here is collector – twitchers are collectors with an impulsive drive to see a bird, in any way possible, be it through a grand strategy or a tactical skirmish. There are birders who like to take their time to see a rare bird or a vagrant (though that may be a fallout of a punishing day job or finances), but twitchers are folks who want instant gratification. How far will that drive take you? Very, very far, as Atul Jain demonstrates in his delightful essay. After all, this is the chap who hopped on to a plane to the Maldives, drove out of the airport, twitched the White Tern and drove straight back to board the same aircraft for the return flight! We don't think the airline crew has recovered from that one!

Twitchers are possessed souls – we act first and think later. It is no wonder that most of us are impulsive, reckless and, sometimes, even compulsive liars. Words like vagrants, rarities, difficult bird, photo lifer and life list excite us. We will go to any length to twitch a new bird – taking leave from work citing the death of a non-existent relative, bribing one's partner with flowers and vacations to exotic locations and, when that doesn't work, cajoling them with folded hands.

How does a twitcher plan to go after a species? I'd be honest in admitting that we are opportunists, often aiming for species that are highly endangered. Commoner lifers – like the White-spotted Fantail for me (yes, unbelievably a lifer) – can wait. It's also important to tick off species that require physical strength, such as the Western Tragopan, while age and strength are still on one's side. And then there is the matter of keeping tabs on vagrants, birds that show up unannounced. My priorities are clear – first go for vagrants (who knows if they will return in our lifetime for a follow-up visit?), then rare birds, followed by other birds. For the first, we rely heavily on information from social media, especially WhatsApp groups (there is even a group for the 'Single Bird Twitcher'!), hobnobbing with bird guides; some of us, who are more technically inclined, set up rare bird alerts on bird listing platforms like eBird.

Once the quarry is scented, the chase begins, and this usually involves cashing accumulated airline miles to drop everything we are doing in the quest of yet another lifer. And then one must choose the right bird guide – you want someone reliable, someone who can show you multiple species. It's simpler when you are chasing a single bird. But for me – I am happy to admit – a guide is a must. I don't claim to be an expert who can find a species that I've never seen before. Take for instance the Yellow-rumped Flycatcher. It took all my persuasive skills to get a knowledgeable local birder to go along with me, and it took him an hour to locate the bird! And then, even after all this scheming and planning, there is no guarantee you'll see the bird. For example, I've made four to five trips to see the Plain Leaf Warbler and am yet to spot it. Plus, there is the honour code – I won't tick a bird unless I have seen it well.

It is not an easy sport and is, most of the time, a solitary one. Every single addition, especially after a certain milestone on the life list, is an arduous task, and requires killer military instincts. Economics plays a role too; many a twitcher has been berated by his or her better half for frittering away money on the quest for a bird. But then, this is how we are structured. As a popular advert goes, '*Kya kare, control nahi hota*! (Sorry, it's out of my control!)'

- The most difficult thing I've ever done to see a bird: trekking in the Great Himalayan National Park to see the Western Tragopan
- The bird I've missed most times: Plain Leaf Warbler
- The longest I have travelled to see a bird (if you see it as a trip to Coimbatore via the United States): Yellow-rumped Flycatcher
- A bird that I have attempted to twitch and failed: Jerdon's Courser

Chin Hills, Mizoram, January 2011

Fifteen years ago, birding the Northeast was not easy. Lack of information due to the absence of social media, bird guides and infrastructure made the birding experience difficult yet exciting. The thrill of going into the unknown and the prospect of discovering a species that was yet to be reported was enough to give me an adrenaline rush.

Three of us – ornithologist Shashank Dalvi; Ramki Sreenivasan (Ramki to the birding world), the late founder of Conservation India and I had been birding together for a couple of years, when in 2011, we decided to explore the Murlen National Park, near the Chin Hills in the Champhai District of Mizoram. The quest was to find the Chin Hills Wren-Babbler (*Spelaeornis oatesi*) and perhaps look for Mrs Hume's Pheasant as well. The wren-babbler was a recent split and there was no photographic evidence of it from India. The idea was exciting, but the trip required months of careful planning. We had to first work out where we were going to look for the bird. Then we had to figure out how we were going to get to the National Park, which is about 250 km from Aizawl's Lengpui airport – roads were rudimentary at best, and it was also hard to find reliable transport and a place to stay. The Mizoram Forest Department was extremely helpful over the phone and fax, and we soon had our permits in hand, but we gathered that the forest rest house was not a viable option to stay at – it had been vandalized and was in ruins.

That's where Ramki's network came in handy. He found out that one of his professors from business school – Natarajan Balkrishnan, who was then teaching at the National University of Singapore – had a household staff member named Sankte who hailed from Champhai, Mizoram's third largest town. This was a happy coincidence as Champhai is right at the base of Murlen National Park. Sankte readily agreed to help us with the logistics, and a plan was hatched. She was the lucky charm who managed to organize our transport and stay for the trip.

Soon, we were making the back-breaking journey from Lengpui airport, which we completed in two stretches. The first part was a twelve-hour drive from the airport, which got us to Sankte's house in Champhai. We were graciously hosted by her family and were fed vegetarian food, which itself was a minor miracle in the remoter locations of the meat-loving north-east. The second part of the journey was short, but on a muddy road that tested all the skills of our driver and the endurance of his vehicle. Nonetheless, we arrived in one piece at our destination, Murlen, and were soon at the doorstep of our host, Munga, who happened to be Sankte's cousin.

Munga was a hardworking farmer who loved fishing – he was an amazing human being. The house was simple and was constructed with wooden planks. It basically consisted of one large, clean, spacious room, which we shared with Munga and his eight family members. The focal point was a wood-fired stove near which we spread our sleeping bags. It was wonderful to experience the hospitality of this simple and large-hearted Mizo family. In the evening, almost all the menfolk of the village turned up to meet and chat with us, with an English teacher doing most of the talking and translating. Staying with Munga's family and experiencing the simplicity of their village life was a humbling experience indeed.

Getting to Murlen was hard; finding the birds wasn't. The next two days were spent in and around the Murlen peak; at 2,000 m, this is the second highest mountain in Mizoram. In no time – within the first five minutes of birding – Shashank easily fished out the Chin Hills Wren-Babbler and the first photographic evidence was established. The bird had popped out from the undergrowth right below us, on a lichen-covered branch, displaying its light brown scaly head and mantle, which contrasted with the darker brown wings and tail. Then it hopped around for a while, and finally perched on a thin branch and turned around to face us, displaying its pale throat, which contrasted with its typical grey ear coverts and the belly dotted with thin black spots. Then it sang, turning left and then right, while craning its neck towards the sky.

Once we spotted the first individual, more followed. On the path that goes up to Murlen peak, we found fifteen individuals of the species within half a day. The mountain also had grassy patches where Mrs Hume's Pheasant was likely to be found; we would need to camp out there to look for the bird – something we didn't have time for on that trip. But the habitat was rich in birdlife. We found Black-crowned Scimitar Babblers, Rusty-fronted Barwings, Blue-winged Minlas, Flavescent Bulbuls, Grey Sibias,

Assam Laughingthrush...typical birds of the North East.

Hunting is a way of life in this part of the world, and we found plenty of snares along the path that had trapped many birds. We cut a few of these and released the trapped birds, doing our bit for conservation; it became clear that hunting in these patches had led to something that is common across the North East – where animals across taxa are hunted for the pot and sport – namely, the 'empty forest syndrome'. Despite this damper, the sheer fun of exploring an unknown area without much infrastructure gave us a high and had us smiling all the way back from our trip.

Narcondam Island, Andaman and Nicobar, March 2011

I personally feel that twitchers are not taken very seriously by wildlife researchers, and are often regarded as casual fun-seekers. On the other hand, the fact remains that most of the data on eBird is being populated by amateur birders, and researchers rely heavily on it. Some birds, however, remain out of reach for the casual birder – one of these is the Narcondam Hornbill. It is found on Narcondam – a remote island that is closer to Myanmar than India and is off limits for civilians. Researchers like Ravi Sankaran had visited it previously to study the endemic Narcondam Hornbill *(Rhyticeros narcondami)*, but, in 2011, we were the first birders to make our way there.

While the Narcondam Hornbill is easy to see once you get to the island, the real challenge lies in getting there! While trying to make plans to land on Narcondam, I knocked on every possible door – the army, the police, the Forest Department, researchers and even divers – but no one was of any help. At that point, I decided to take matters into my own hands and fished out the contact number of a certain Nick Band – a sailboat owner, based out of Thailand. He was conducting fishing tours in the Andaman seas and could potentially take us to the Island or at least nearabout. Nick agreed to bring his boat to Port Blair (from Thailand) and was able to wrangle a permit for angling around Narcondam Island. It was a different matter altogether of how complicated it was to get the payment across to him before he agreed to set sail from Thailand.

Five of us – Harkirat, Manoj, Vinay, Aparna and I – boarded the fishing boat at Port Blair and were soon sailing with two crew members, namely, Nick and his Thai partner. It was a small boat which could sleep only two people in the cabin. This privilege was extended to the only couple, Aparna and I; the rest had to rough it out by sleeping on the deck. Our sail speed was around 4–5 knots/hour, which was rather slow, but the progress was enlivened when Nick gave us quick sailing lessons so that he could catch a few winks of sleep.

There were some funny, anxious and 'wow' moments on this approximately twelve-hour journey. Manoj was answering the nature's call when shouts went up in the air – pelagic birds! He rushed to the deck with the job unfinished, but the birds were gone by the time he arrived. Excitement spread when we sailed past Barren Island, India's sole active volcano (which was active at the time). We could see lava pouring into the sea as well as smoke and ash spewing out of the crater's mouth. Somewhere around Barren Island, a scare was raised when our fathometer – an instrument used to measure ocean depth – started showing 'zero', a sure sign that we were heading for a crash against a submerged sea island. The earth's tectonic plates had changed after the 2004 tsunami and, thus, nautical charts were not accurate. Nick was promptly woken up, but his anxiousness didn't help us at all. However, after a few minutes, he declared that the instrument was faulty. This further added to the sense of chaos.

Meals were the catch of the day, and rice for breakfast, lunch and dinner. The drinking water was heavily rationed, and bathing was forbidden. The last rule triggered a scene as Nick thought Aparna had indulged in the luxury of a bath, which meant there would be no water left for the return journey. After weathering these minor storms, we arrived at Narcondam around dusk and docked not far from the Island. Made of volcanic rock sprouting from the sea in the middle of nowhere, like a pyramid, the island looked tiny and seemed around a kilometre from tip-to-tip. Through our binoculars, we could see birds flying and were easily able to spot the hornbill. In fact, the Narcondam Hornbill was the most common bird on the island. The male hornbills had rufous necks with white throats, which contrasted with their black bodies. The female hornbills, on the other hand, were completely black with a contrasting white throat. Both sexes showed prominent ridges on their casque which signalled the age of the bird. This species has defied all logic and science by managing to colonize this tiny oceanic island in the middle of nowhere. Worldwide, the data shows that the smaller and further an island is from the mainland, the more minuscule the chances of birds colonizing such islands. However, somehow, the Narcondam Hornbill has managed to reach, and even thrive, on this remote island.

Narcondam Island has a small police presence, and we were paid a visit by this team the following morning. This is a remote posting, and the change of guard takes place every month. Devoid of external human contact for weeks, our presence was a welcome change for those policemen. We were welcomed aboard their inflatable boat and were soon cruising towards Narcondam. Two policemen doubled up as our guides and gave us a quick tour of the Island, which appeared to support a stable population of hornbills. Our hosts also extended us the hospitality of a bath, fresh food and water refills; the latter was carried back to our boat in drums. Nick was mighty relieved at this gesture! The return journey was uneventful, except for a stopover at Neil Island to watch many Andaman endemics.

Eaglenest Wildlife Sanctuary, Arunachal Pradesh, May 2009

Twitching is all about planning, but sometimes the birds will have none of it. I had already made two trips to the Eaglenest Wildlife Sanctuary in Arunachal with Shashank, before the lure of the Hodgson's Frogmouth took me there again. Shashank had worked with Dr Ramana Athreya in this wildlife sanctuary and was part of the core team that mapped out the Bugun Liocichla's distribution in western Arunachal Pradesh – his intimacy with the birds in this forest was unparalleled. Until the 2000s, no one had seen the Hodgson's Frogmouth *(Batrachostomus hodgsoni)* in India for a long time (one had to travel to Bhutan to spot it, but even there, there had been just one sighting of the bird!). But Shashank had promised to try and find this bird for me in India, and in 2009, while birding in Eaglenest in May, he spotted the frogmouth on his very first attempt, once he figured out the right vocalization of the species. When he called to inform me that he had found the frogmouth in Eaglenest, I dropped everything and flew to Guwahati, even though I was already booked to join a tour led by Shashank in the very same place just ten days later. A twitcher's life is full of uncertainties and dreaded 'what ifs'. In this case, what if the bird disappeared from India again?

It was election time in Assam then, and unoccupied cabs, even if they were en route to collect a passenger, were commonly commandeered for election duty, with no choice offered to the driver or the hapless passenger waiting to be picked up. As a result, Shashank drove all the way with Raju, our driver, to collect me from Guwahati, and then we decided to enter Eaglenest by the shorter route from Doimara, through the southern gate of the sanctuary. This made sense since he had seen the frogmouth in the southern region of the sanctuary.

Doimara lies at the lowest elevation of Eaglenest. The historic Foothill-Chaku-Tenga road (more like a track!)

takes travellers from Doimara to the highest point in that region, which is Eaglenest Pass, before dropping to Lama Camp, which is the base camp to find the Bugun Liocichla. Just past the Doimara bridge is a zone that is prone to landslides after rainfall. That day, soon after we crossed the bridge, as we were driving up, we found a landslide blocking the road. It had to be cleared manually – luckily, the landslide mainly comprised soft white sand, and removing it wasn't too difficult. By the time Raju accomplished this task, it was already dark and it had started raining. Soon, the debris slowly started to come down again! The vehicle skidded and came under the heavy firing of rubble falling off the mountainside. We would have been pushed into the river below, had it not been for the driving abilities of Raju who kept his cool. The vehicle was still in the middle of the landslide zone when it came to an abrupt halt, and Raju couldn't restart it! Shashank and I then tried to push the vehicle up through the part of the track that had been cleared, even as the rubble continued to fall. It was minutes (thought it felt like hours!) before Raju miraculously managed to start the vehicle and get it across the landslide, and on to the road on the other side.

By the time we reached the Khellong checkpost, it was pouring; we had heard at least three different frogmouth individuals by then, but we were able to bird only intermittently, between spells of heavy rain. A lot of elephant activity was also prevalent in the area, so it wasn't safe to go too far from the car. Darkness, rain, landslide, elephants ... but no frogmouth in sight yet. We were dozing in the car, waiting for yet another shower to stop, when we had a 'Jatinga'-like experience.[1] A thud on the jeep window woke us up with a start. We saw that a Silver-eared Mesia had become disoriented and had flown into our stationary vehicle, before recovering from the shock and flying away to safety. That start in the dead of the night, on the heels of a perilous escape from the landslide, left us even more petrified. Once equanimity had been regained, we noticed that the rain had come to a halt; it was time to look for the bird on foot. We trudged along for a few more hours before we decided to give up our search and head to the camp at Sessni which was higher up the hill.

The second part of this story was revealed many years later. While we were walking around looking for the bird, Raju had opened the jeep's windows as it was getting stuffy. He also urgently needed to answer the nature's call but was too wary of the nearby elephants to step out of the vehicle. The open window attracted a bat, which began circling inside the jeep, near the roof. That was the last straw for the poor man, who promptly peed in his pants out of fear. Raju, who was brave enough to counter a raging landslide, had been pinned by a harmless bat!

The next morning, rather than head straight to the frogmouth patch, we decided to bird around Sessni. It was an amazing morning, and we ticked off Pale Blue Flycatcher, Bluewinged Laughingthrush, Sikkim Wedge-billed Babbler, Eyebrowed Wren Babbler, Golden Babbler and many other birds typical of the middle elevations of

Eaglenest. Post lunch we drove back towards Khellong; the sky being overcast, Shashank had a hunch that the frogmouth, a nocturnal bird, might reveal itself during the day.

A three-hour search ensued as we followed the calls of the bird along slippery bamboo slopes. We saw some amazing birds along the way – the Sapphire Flycatcher, the Yellow-bellied Warbler, the Pale-headed Woodpecker, the White-hooded Babbler – before we found a female Hodgson's Frogmouth, sitting perfectly camouflaged rather unusually on a non-bamboo patch within the larger bamboo-covered area. The bird looked sleepy and didn't pay attention to us birders, slipping and falling as we were in our eagerness to catch a glimpse of it. In the morning *'darshan'*, the frogmouth sported a look that was quite different from its night 'attire', when they display distinct hairy feathers on the head (ironically, these feathers protect them from their prey species, which are primarily insects). During the day, these feathers were flattened atop the head and were impossible to see.

Seeing a nocturnal species during the day is always amazing. Shashank even found one near Ramalingam (in the northern parts of the sanctuary) just ten days later, when I came back for the scheduled tour. In the fifteen years since, the Hodgson's Frogmouth has been found at several sites in northeastern India.

Little Andamans, Andaman and Nicobar Islands, December, 2018

Six years ago (and perhaps even now), the Mandarin Duck *(Aix galericulata)* was an enigma for Indian birders. A striking-looking duck with an orange face, red beak and a purple breast that contrasts with the orange sail fin feathers on its back, this species is found across East Asia. There had been a few sporadic sightings of the duck in Northeast India, but when the news came that the bird was being seen in Little Andaman, and with consummate ease, I knew it was my chance to tick off the species.

December is not a great month to travel to Port Blair, for all flights and hotels are full of holidaymakers. Undeterred, I called up Vikram Shil, a local bird guide, and requested his help. He informed me that I was in luck, as a group of his clients from Pune were leaving for Little Andaman the next night; he offered to book a ticket for me along with them on an overnight ship and suggested I bird with them. I readily took him up on his offer and bought an expensive ticket to Port Blair; at those ticket prices, I could have vacationed in Singapore!

I met my co-birders at the harbour on the evening of our departure, and we made our way to the ship. Warned by Vikram that sleeping in the ship's cabins would not be a pleasant experience, we cornered a spot on the deck and spread our bedsheets to stake claim over our respective 'sleeping quarters'. The balmy breeze and the star-lit sky led to lots of birding banter over a few drinks and packed dinners. We slept fitfully as the deck was hard on our backs; by eight the next morning, we arrived at Little Andaman.

My original plan was to spend a night on the island before taking a fast boat back to Port Blair the next day to catch the return flight to Delhi. But the previous evening, while on board, I had heard to my dismay that the fast boat for the return journey the next day had been cancelled, as it had been pressed into the service of a visiting dignitary. This created a lot of anxiety, as there was no other means of travelling back to Port Blair, except on the very same ship that was sailing back within an hour of its arrival. Delaying my return to Port Blair not being a possibility, I had to find the bird within that hour; and so, everything hinged on being the first to exit the ship, finding the vehicle that would take me around quickly and reaching the spot that the duck was frequenting without any delay. My co-birders had no such pressing need, as they were staying for a few nights on Little Andaman.

With much at stake, I used every last ounce of my sales skills to persuade the captain of the ship. I can say with conviction that he became a convert to my cause and was baptized into birding that very night. He agreed to delay the boat's departure by 30 minutes and also to put us at the head of the queue of deboarding passengers. Once the boat docked the next morning, Vikram's group and I jumped into our vehicle and made our way to the Mandarin Duck spot. To our chagrin the bird was not far (all that anxiety over a wild chase, with the clock ticking, that had been playing in my mind all morning was clearly for nothing), and was sitting comfortably on the fringes of a small pond, oblivious to our exertions on its behalf. It was a stunning-looking colourful featherball, a male with a red-and-white bill. The green–purple colours of the crown contrasted with the bird's massive white supercilium, while two distinct white bars on the side of the breast also stood out. What made the sighting even more memorable was the surroundings, as the bird was in the middle of a colourful lily (false lotus) pond. I watched it for some time, and by the time I was done, there was still time for breakfast – excellent dosa can be had at Santosh Bhojanalaya in case anyone ventures to Little Andaman – before climbing aboard the ship to sail back to Port Blair.

It was the twitch with the smallest ratio of time spent with the bird (not even 10 minutes) to overall journey time, but was totally worth it!

Mandala, Arunachal Pradesh, April 2023

Not all twitches are successful. That makes a successful rematch all the more sweeter. Three years ago, I had dipped on the Temminck's Tragopan *(Tragopan temminckii)*, and in frustration booked a hotel one year in advance for a rematch. I was determined to simply camp at a spot favoured by the bird and make multiple attempts to see it. Like others in its family, the Temminck's Tragopan is an exquisite-looking bird, with an orange body that vividly contrasts with its electric blue facial wattles. It is found in the mountain forests of Northeast India and is high on the target list of birdwatchers and photographers alike.

The second time around, the key strategy was to try for the bird in a more suitable season – in spring – when the birds are pairing up and hence more vocal. Perhaps I would get lucky then. I requested Aparna to come with me so that I could introduce her to the beauty of Arunachal. She agreed and it was on 1 April that we reached our hotel in Dirang, a town conveniently located for birding in the Mandala region that hosts the tragopan.

I was pleasantly surprised to find that Peter Lobo was also staying at the same hotel. Peter is one of India's finest bird tour leaders, and has been quoted in many birding accounts and books for his extraordinary birding skills. He was then leading a group of Swedish birders, and when I asked him if our vehicle could follow his, he readily agreed, and we left early the next morning for Mandala. The first stop was a very well-known tragopan location. We walked down the trail that passes through a ravine and tried calling in the bird, which did not respond, and we could not see any signs of its presence. Peter and his group moved on to scout another location a few kilometres down the road. While Aparna and I were walking back up the trail, we saw a flock of Crimson-browed Finches, a species new for both of us. We scanned the spot for other species for another 15 minutes before moving back to the car and driving ahead. Our driver was driving slowly as instructed, and soon after crossing an army camp, we saw a bird crossing the road. The Temminck's Tragopan! We just couldn't believe our luck, and the word serendipity found a new meaning in our lives. We played the call and the bird came back again in all its regal glory. It was a blazing red football with white ocillies or spots on the back and belly. Its black face contrasted with blue facial skin and the golden yellow crown feather. At that moment it was difficult to imagine any bird more beautiful than this individual. We marked the spot and drove ahead to meet Peter and his group of birders. We explained the exact location, which helped Peter fish out the tragopan easily for his group.

The cold breakfast that day tasted like gourmet food! I had thrown all planning aside and decided to simply be in the right place at the right time to make the sighting happen. It did, and the sight of the tragopan highlighted a day that will remain forever etched in our memories.

Jessami, Manipur, October 2022

It is sometimes nice to be a second choice. When I got a call from Sudeshna, a twitcher par excellence, to go look for a long-coveted species, my only impulse was to say yes. Sometimes you just have to say yes and commit to a twitch. Then you worry about how to make it happen. In this case, Sudeshna's first choice had backed out from the trip, and I filled the spot without even thinking about professional and personal commitments scheduled during the travel period; I had lined up five conference calls for those days, of which four I postponed and one I just forgot!

The bird in question was the Mrs Hume's Pheasant (*Syrmaticus humiae*) – a mythical species in India until recently – when a credible location was established at a village called Jessami in Manipur. Hereto, this bird was almost impossible to see in India due to widespread hunting and a huge decline in its population.

It was the first week of October 2022 when Sudeshna and I reached our homestay in Imphal for an overnight stay, before leaving early the following morning for Jessami, with our guide Harmenn Huidrom. The eleven-hour drive was long and treacherous, but ultimately uneventful. Late that day, we pulled up at a community centre, which was managed and run by the Jessami Village Council, and which was to be our base for the next two days. We were greeted by the village council's chairman and his colleagues and shown to our dormitory. A quick dinner followed, and then we were ready for the birding action, which played out with military precision.

Twelve able-bodied men were chosen to accompany us, in vehicles arranged by the council chairman. We travelled for about 45 minutes in the dead of night before the vehicles rolled to a stop. Two trekking parties were formed; one went left and the other right, while we waited in our vehicle for a 'positive signal'. The wait stretched for more than three hours, and at around 1.00 a.m. we saw a flicker of light 'walking' towards us from the left side. All possible gods were prayed to, and blessings summoned, with the hope that it would be a thumbs up. It indeed was, and we were asked to follow the man who had returned from the jungle. We trudged along with him for a couple of kilometres inside the forest until we came upon the rest of the party, who were sitting on the forest floor and chatting casually. They all wanted to come with us; however, we asked Harmenn to request them not to accompany us lest the bird get disturbed. So, only the three of us followed the spotter further downhill until we were signalled to stop and sit. Our hearts were pumping with excitement. And then one of the finders pointed a torch at a tree to illuminate a family of Mrs Hume's Pheasants – a male, a female and a juvenile – roosting in the mid-canopy. What a glorious and beautiful-looking bird! The male looked resplendent, with its chestnut chest and a long greyish tail. The female was no less attractive, with black dots on its upper body, though it was much smaller in size than the male.

High fives were exchanged, and we profusely thanked the team that had implemented the search with such precision. Familiar as they are with the forests, and possessing the ability to tread without the slightest sounds that could cause the much-hunted pheasant to take flight, teams of hunters working in concert to scan a hillside at night have become indispensable to the sighting of these birds. It's hard to escape the irony that the same hunters who have taught the bird to become invisible in the forests of the Northeast are now helping birders tick this species off their lists.

Sholayur, Palakkad Kerala, January 2024

The Yellow-rumped Flycatcher (*Ficedula zanthopygia*), a bird of eastern Asia and a vagrant to India, is one of my latest scores. It has limited records from India, and twitchers consider it one of the holy grails for India. In early 2024, a female bird kept showing up for more than two weeks near Coimbatore; the bird had been ably located by a group of birders from Kerala. The thought of twitching it caused me to toss and turn in bed at night, as work commitments required immediate travel to the United States. Even as I reached the airport to board the flight, my constant thought was to somehow find a reason not to go, but I could not produce any credible explanation that wouldn't also jeopardize my career. My return after a week, heavily jet lagged after a nineteen-hour flight, was no deterrent. I jumped into another plane headed to Coimbatore within a few hours of reaching Delhi. It did help the cause that my wife was also travelling for work – one less problem to solve!

This particular twitch underscores the importance of the 'network' for all birders, not just for twitchers. I was boarding a flight with very little by way of preparation to find the flycatcher. A few quick calls to birding friends, specially to Praveen J., the eminent editor of the *Indian Birds* journal and a brilliant birder besides, were made and a local birding pal, Karthik, was identified to accompany me the next morning. Having time on my hand before picking up Karthik, I decided to head out to the legendary tiffin outfit – Annapoorna Gworishankar. Soon enough, wheat upma and a ghee roast dosa were on the table and were quickly devoured accompanied by an equally amazing cup of filter coffee.

Fortified, I picked up Karthik and we headed out to the Anaikatti range forest located at the foothills of the Nilgiri Biosphere Reserve. Once we reached the spot frequented by the bird, my heart rate increased with trepidation – was this going to be a 'duck' or a 'tick'? We scanned all possible trees and shrubs but there was no sign of the bird. It took nearly an hour of intense searching before Kartik spotted the bird in the mid-canopy of a large tree. Once we got the target in the lenses of our binoculars, we tracked the bird as it hopped from one tree to another. This much sought after bird, being a female of the species, was visually unimpressive, being dull brown with a long, white wing patch and a yellowish belly. Nonetheless, lifer number 1,215 was in the bag!

I am often asked by my family, friends and sometimes by fellow birders, what the point of twitching is. It is difficult to respond, as there is no right answer. Hunter-gatherer instincts? A one-up-man-ship sport or a pure ego trip? Hard to say. I feel that the answer probably lies closer to the desire to find an excuse to be in the wild, topped with an addiction to the adrenaline rush that comes with every new find and a desire to be on top of the game. My trips have taken me to the remotest corners of India and helped me experience different cultures, food and people. It has also made me mentally tough, teaching me to take success and failure in my stride and to never ever give up trying. This hobby has even helped me find a group of companions who will be friends for life.

The economics of twitching have changed substantially since the time I took up the sport. It also has to do with the stage one is at. The first 700 birds are not so challenging. From 700 to 1,000, it is a little difficult as you have to engage experts to find the birds. After 1,000, the cost component increases exponentially. Once your list crosses 1,100 species, every single bird costs quite a lot of money because you are spending on flights, hotels and food, all for a single bird. But that's my passion and I'd rather spend money on this than buy a fancy car.

What does it take to be a successful twitcher? I'd say that you need to be persistent, have a sense of adventure, network like crazy and possess the courage to say 'yes' when the opportunity presents itself. It also helps to have sharp sales and people skills, which come in handy when convincing sundry government officials, office colleagues and, of course, the spouse, to let you make that trip. Having an understanding partner who gives you a hard time for every single crazy, last-minute trip but always relents in the end is priceless. With all this on my side, I have many reasons to 'Keep walking'!

ABOUT THE AUTHOR

Atul Jain has been birding for the last 24 years, a sport he got into by accident. On a BNHS trip to Dudhwa National Park (Lakhimpur District, Uttar Pradesh) in 1999 (to watch tigers, what else!), he happened to see an Emerald Dove. And he was smitten! The resplendent colours, beauty and shyness of the bird enchanted him no end, and in that moment, he became a birdwatcher. When this turned into a chase game for numbers (twitching) is not known, but it seems that this bug had always been there. It is reflected in his collections of bird memorabilia, banknotes, stamps, lithographs, paintings and even betel nut crackers among many other sundry items. Atul thinks that if not for birding, he wouldn't have had the chance to visit the remote places in India and experience the magic of forests, wildlife, culture and food. The sheer adrenalin flow of visiting a hitherto unknown territory is what sends him reeling from one flight to the next. Atul works for a multinational company. He is married and has a daughter. He is an avid foodie and lives in New Delhi.

12

Finding the Next Rare Bird for India

FRANK E. RHEINDT

Having come this far, the question for birders and biologists is – what is next? It is time to look ahead to find out what else can be unearthed in India and what geographies to venture forth to in search of a rare bird. The question also for most of us is – what's on the list? Frank Rheindt, who has discovered/re-discovered a number of species across Southeast Asia has some interesting answers. Be warned though – the new puzzles for Indian ornithology will be the toughest to unlock. This essay, however, has some of the keys.

Covering almost 3.3 million sq. km at mostly tropical and subtropical latitudes, India is one of the most biodiverse countries in the world, ranking first among all nations in terms of the number of vascular plant species, and ninth in terms of birds. The country forms its own subcontinent, although important parts of India spill over into neighbouring zoogeographic zones, especially the Andaman and Nicobar Islands and the northeastern hill states, whose fauna is more Southeast Asian in character. By all accounts, it is extremely unlikely that our current knowledge of India's birds is complete. In such a vast and diverse country, new discoveries are certain to lurk beyond the horizon.

At the same time, India's biological riches, including its birds, have been comparatively well explored. The vast bulk of its bird species was described to science on the basis of the collections of a few indefatigable naturalist-adventurers, such as Edward Blyth (1810–1873) and Allan Octavian Hume (1829–1912), during the golden era of biological exploration in the late nineteenth and early twentieth centuries. They laid the foundation for the work of Sálim Moizuddin Abdul Ali (1896–1987), the Father of Indian Ornithology, who, together with Sidney Dillon Ripley (1913–2001), consolidated our knowledge of India's birds in their landmark *Handbook of the Birds of India and Pakistan*.

Standing on the shoulders of these giants, today's ornithologists and birders in India have the benefit of an unusually detailed account of their country's avifauna, which goes far beyond our knowledge of birds in many other tropical regions. For instance, in this new millennium, only three bird species that are known to range in India have been described as new to science: Ashambu Sholakili, Himalayan Forest Thrush and Bugun Liocichla. In contrast, the number of new bird species described during this period is roughly ten times higher for Indonesia, another tropical Asian country, which is only roughly half the size of India. This low incidence of new discoveries attests to the comprehensive and thorough ornithological work that Sálim Ali and his predecessors carried out across India.

Many future detections of bird species new to India will doubtless be serendipitous and will be made in moments when the discoverer least expects it. However, the discerning birder may increase her chances of making exciting discoveries by focusing on areas and strategies that appear most promising. Such strategies are largely based on our knowledge of biogeography, climate and patterns of bird migration. Here are a few such ideas that birders may find helpful.

Pelagic Waters – The Great Unknown

The coastlines of India are vast, and its island possessions (e.g., Lakshadweep, Andaman and Nicobar) provide the intrepid birder with access deep into the Indian Ocean. Pelagic birds, including such enigmatic groups as shearwaters, petrels and storm-petrels, critically rely on cold-water upwellings in deep sea that provide nutrients for phytoplankton, which in turn attracts krill, fish and other food species. Yet India's deep seas remain largely unexplored from an ornithological perspective. This is unfortunate because India's coastline is particularly conducive to pelagic bird observations. In contrast to most coastlines in other large Asian nations (e.g., China, Indonesia), India's coastlines drop precipitously, making excursions into deep-sea waters particularly easy. For instance, from Chennai, it only takes a 50 km boat ride to reach waters of truly oceanic depth, going as deep as 3,000 m below the surface of the sea – a geographic feature that few other Asian cities share in common. Keen observers are encouraged to undertake frequent pelagic expeditions into the oceanic waters of the Arabian Sea, especially off the coast of Kerala, Karnataka, Goa and Maharashtra, or explore the seas around Lakshadweep. That part of the Indian Ocean is particularly promising, as it adjoins areas of high pelagic bird endemism around Socotra, the southern Arabian coast, the Seychelles and Comoros, with a chance of some of the petrels, shearwaters and other pelagic species of those regions turning up in Indian waters.

The Eastern Palaearctic Flyway

India receives millions of migratory birds every winter, and serves as a stopover point on migration for millions more. Its geographic position is right in the centre of the winter quarters along the Central Palaearctic Flyway, which connects the birds breeding in Siberia and Central Asia with their non-breeding range in South Asia and the Middle East. Two other flyways flank India to the west and the east. One of them, the Western Palaearctic Flyway, mainly connects Europe with Africa. However, since many European species range far into Siberia and brush past northwestern India on migration, Indian birders have long known to look out for Western Palaearctic specialties (e.g., Spotted Flycatchers) on migration in dry border states such as Gujarat. What has largely been neglected, however, is India's exposure to the occasional Eastern Palaearctic specialties in the Andaman and Nicobar Islands, and the northeastern hill states (e.g., Nagaland, Arunachal Pradesh, Manipur, etc.).

The Eastern Palaearctic Flyway connects the species breeding in Japan, Korea, far eastern Russia and eastern China with their Southeast Asian winter quarters. Given the generally lower level of ornithological exploration of Northeast India, even common Eastern Palaearctic migrants have long been entirely neglected; for instance, Eastern Yellow Wagtails and Two-barred Warblers are almost guaranteed to be common winter visitors in some areas of Northeast India, even more common than their western counterparts (i.e., Western Yellow Wagtail and Greenish Warbler). And yet, it took until quite recently for them to be regularly noticed. Other Far Eastern migrants have also been recently detected for the first time in India (e.g., Eastern Crowned Warbler, Band-bellied Crake). How many other species from the Far East have been overlooked in Northeast India or the Andamans during the winter? For the discerning birder, India's first Radde's Warbler may be lurking just around the corner during their next trip to Namdapha (Arunachal Pradesh).

The Missing Burmese Element

India and Myanmar are divided along one of the most biodiverse border regions on Earth. The hills and mountains, stretching from Arunachal in the north to Mizoram in the south, have a distinct endemic element, and harbour birds from flat grasslands and monsoon woodlands across hill forest, and all the way to high-elevation treeline habitat. The mountains on both sides of this border have been subject to a comparatively thorough level of historic ornithological exploration, and few new species have been described from here. Even so, discoveries of bird species new to India could be expected from two distinct sources.

For one, Myanmar has the larger share of this mountain range, and most of the higher mountains are on Myanmar's side, bar those that are right at the borderline. This means that a few montane species have historically only been known from Myanmar, where they are guaranteed to be more common. Birders keen to look for these species are encouraged to explore the highest elevations of Mizoram, Manipur and Nagaland. One mountain that has featured prominently in the history of ornithological exploration in this area is Mount Victoria, in the Chin Hills of Myanmar. This mountain forms a centre of bird endemism, harbouring disjunct populations and even species that are almost entirely restricted to it. One of its near-endemics, Mount Victoria Babax, was known from India only from historic collections, and it took a long time until it was rediscovered on the Indian side of the border, in Mizoram. Victoria

Nuthatch (or the White-browed Nuthatch) may never be found in India, as neighbouring Mizoram doesn't have mountains tall enough for its preferred habitat zone. But other montane species of these border mountains, such as Yunnan Fulvetta, are known from spots in Myanmar that are only a few kilometres beyond the border with Arunachal and Nagaland, and are just waiting to be discovered on the Indian side.

Further down from the higher elevations, each country has its own complement of lowland avifauna, with ample sharing. However, there are reasons to believe that Myanmar-specific lowland birds may soon be found on the Indian side, especially along Manipur's eastern border with Myanmar, where the Indian territory reaches down the eastern slopes of the border range and abuts the Burmese lowlands along the Chindwin River. While most of these areas are remote and tribal, some of them can nowadays be conveniently reached by road, around the Moreh border crossing in Manipur. In time, lowland species of open habitat, such as Burmese Bushlark or Burmese Myna, may be found on the Indian side with an opening of the landscape. The Indian parts of these lowlands would previously have been covered by dense monsoon woodland and forest, but habitat conversion and even a changing climate – with more frequent dry spells and droughts – are likely to lead to the availability of grasslands and edge habitat that may soon be invaded by species specialized in the open landscapes around Myanmar's large rivers. Excursions to these areas would certainly have a decent potential to turn up new lowland species for the list of India's birds.

Birdwatching is now a blossoming pastime across India. The further growth of field activity across the country, coupled with citizen science platforms, such as Xeno-canto, eBird, iNaturalist and others, is certain to produce many exciting new additions to India's bird list. Observers are encouraged to venture off the beaten path and keep an open mind for any unusual-looking bird they may find. The search for India's rarest birds is indeed still on!

ABOUT THE AUTHOR

Frank Rheindt is a professor and researcher at the National University of Singapore. He is a passionate global birdwatcher who has seen over 9300 species worldwide. He has spent a lot of time exploring remote areas of tropical Asia, where he has discovered, co-discovered and/or scientifically described roughly a dozen new bird species and half a dozen additional subspecies. In his lab, he focuses on using genomic and bioacoustic tools to unravel the strategies by which birds diversify across the tropics. His publications have led to the recognition of many dozens of splits that used to be merged under larger umbrella species, with important implications for bird conservation.

Notes

1. A Charismatic Duck Painted in Carnation and Chocolate

1. Grewal, B., 2014. ['This is supposedly the first painting ever done of the Pink-headed Duck. Painted, in 1777, by Musavir Bhawani Das ...'] Facebook posts of 5 August 2014 and 17 March 2020. Webpage URL: https://www.facebook.com/share/wwU5RR5sWnMXMMLh/?mibextid=WiMSqg.
2. Fisher, C. & Kear, J., 2002. The taxonomic importance of two early paintings of the Pink-headed Duck *Rhodonessa caryophyllacea* (Latham 1790). *Bulletin of the British Ornithologists' Club* 122 (4): 244–248.
3. Topsfield, A., 2019. The natural history paintings of Shaikh Zain ud-Din, Bhawani Das and Ram Das. In: Dalrymple, W., (Ed.). *Forgotten Masters: Indian Painting for the East India Company*. Philip Wilson Publishers, Great Britain. pp. 40–75.
4. Topsfield, A., 2019. *op. cit.*, p. 41.
5. Kumar, N., (Ed.) 1971. Image of Patna (A supplement to Patna District Gazetteer, 1970). *Gazetteer of India: Bihar*. Government of Bihar, Patna, India. pp. i–xiii, 1–214+29 p.ll. [p. 37: '(various shades of red from pink to carmine) was prepared from shellack ...']

6. Archer, M., 1948. *Patna Painting*. 2nd ed. David Marlowe Ltd, England, for The Royal India Society. pp. i–xv, 1–47+45 unnumbered plates.
7. Art UK, 2023. https://artuk.org/discover/artists/das-bhawani-active-17771782 (accessed on 27 November 2023).
8. Fisher, C. & Kear, J., 2002. *op. cit.* p. 244.
9. *ibid.*
10. *ibid.*
11. Fisher, C. & Kear, J., 2002. *op. cit.* p. 246.
12. Sawyer, F.C., 1949. Notes on some original drawings of birds used by Dr. John Latham. *Journal of the Society for the Bibliography of Natural History* 2 (5): 173–180. [Sawyer misses listing Nathaniel Middleton's painting collection!]
13. Latham, J., 1787. *Supplement to the General Synopsis of Birds*. 1st ed. Leigh & Sotheby, London. pp. i–iii, 1–298 (pl. CXIX). ["Pink-headed D.[uck]"; "Oude [=Awadh]" "Is often kept tame." (pp. 276–277).]
14. Latham, J., 1790. *Index Ornithologicus, sive Systema ornithologiæ; complectens avium divisionem in classes, ordines, genera, species, ipsarumque varietates: adjectis synonymis, locis, descriptionibus.* 1st ed. Leigh & Sotheby, London. Vol. II of 2 vols. pp. 467–920 [p. 866].
15. Fisher, C. & Kear, J., 2002. *op. cit.*, p. 247.
16. In a rare error, the Swedish naturalist, Carl Jakob Sundevall, misidentified the Lesser Whistling Duck *Dendrocygna javanica*, as Latham's *Anas caryophyllacea*. Sundevall, C.J., 1847. The birds of Calcutta collected and described by Carl J. Sundevall. *Annals and Magazine of Natural History* 19 (125): 164–173.
17. Latham, J., 1787. *op. cit.* pp. 276–277.
18. Bowyer, T.H., 2004. Middleton, Nathaniel (1750–1807). In: Matthew, H.C.G. & Harrison, B., (Eds) *Oxford Dictionary of National Biography*. Oxford University Press, Oxford. 38: pp.73–75.

19. Stephens, J.F., 1824. *General Zoology, or Systematic Natural History*. Thomas Davison, London. Vol. 12 (2) of 14 vols. pp. [i–vii], 1–264. ["Pink-headed Pochard *Fuligula caryophyllacea*", p. 207.]

20. Eyton, T.C., 1838. *A Monograph of the Anatidae, or Duck Tribe*. 1st ed. Longman: London. pp. [i–viii], 1–178+5 (p. 152).

21. Prestwich, A.A., 1974. The Pink-headed Duck (*Rhodonessa caryophyllacea*) in the wild and in captivity. *The Avicultural Magazine* 80 (2): 47–52 [p. 48: "*Rhodon*, Gr., a rose, hence red; *esson*, Gr., less or weaker = rose-tinted. *Caryophyll*, Lat., Pink (*Dianthus*); *acea*, Lat. suffix meaning of or pertaining to."]

22. Reichenbach, H.G.L., [1853]. *Handbuch der specielen Ornithologie. Die Vogel*. The Zoological Museum of Dresden, Dresden and Leipzig, Germany. pp. i–vii, 1–36, I–XXXI.

23. Ali, S., 1960. The Pink-headed Duck *Rhodonessa caryophyllacea* (Latham). *Wildfowl Trust 11th Annual Report*. 1958–1959: 55–60.

24. Chagnoux S. & Callou C. 2024. The birds collection (ZO) of the Muséum national d'Histoire naturelle (MNHN - Paris). Version 42.341. https://doi.org/10.15468/h0xtwv (accessed via GBIF.org on 13 January 2024). https://www.gbif.org/occurrence/1042802933 (accessed on 12 January 2024.) [The webpage also contains a collection of breath-taking multi-angle photographs of the mounted bird.]

25. Rookmaaker, K., 2019. Mauled by a rhinoceros: The final years of Alfred Duvaucel (1793–1824) in India. *Zoosystema* 41 (14): 259–267. https://doi.org/10.5252/zoosystema2019v41a14.

26. Warr, F.E., 1996. *Manuscripts and drawings in the Ornithology and Rothschild Libraries of The Natural History Museum at Tring*. 1st ed. British Ornithologists' Club, Hertfordshire, in association with The Natural History Museum. pp. i–xiv, 1–100. ['A keen sportsman and naturalist, he built a museum at Eyton Hall near

Wellington and amassed a large collection of birds and bird skeletons. When the collection was sold in 1881, some of the more important skins and skeletons were selected by Dr R.B. Sharpe for The Natural History Museum.' (p. 27).]

27. Eyton, T.C., 1838. *op. cit.* p. 152.
28. Martin, K., 2021. Colonel Thomas Alexander Cobbe. https://www.findagrave.com/memorial/225686604/thomas-alexander-cobbe (accessed on 12 January 2023.)
29. Thomas, O., 1906. Mammals. In: British Museum (Natural History) (ed.). *The History of the Collections Contained in the Natural History Departments of the British Museum. Vol. II. Separate Historical Accounts of the Several Collections Included in the Department of Zoology*. British Museum (Natural History), London. Vol. II of III vols. pp. 1–66 [6].
30. Cotton, E., 1927. Editor's note book. *Bengal Past & Present*, 34 (Part 1 No. 67): 63–75 (73). https://archive.org/stream/in.ernet.dli.2015.32673/2015.32673.Bengal-Past-And-Present--Vol33-34_djvu.txt#:~:text=Colonel%20Thomas%20Alexander%20Cobbe%20(1788%2D1836)%20who%20was,to%201836%2C%20married%2C%20according%20to%20Major%20Hodson (accessed on 12 January 2024.)
31. Salvadori, T., 1895. *Catalogue of the Chenomorphæ (Palamedeæ, Phoenicopteri, Anseres), Crypturi, and Ratitæ in the Collection of the British Museum*. British Museum of Natural History, London. Vol. XXVII of 27 vols. pp. i–xv, 1–636 [63].
32. Gray, G.R., 1844. *List of the Specimens of Birds in the collection of the British Museum. Gallinæ, Grallæ, and Anseres. Part III*. British Museum, London. pp. 1–209. [Lists three specimens. "*Anas caryophyllacea*, Lath. *Ind.* II. 866. *Callichen caryophyllaceum*, Eyton. (a). Northern India. – From Col. Cobb's (sic) Collection. (b). Bhotan. – Presented by the Hon. East India Company. (c). Female? Nepaul. – Presented by B. H. Hodgson, Esq." (p. 136).]

33. Ali, S., 1960. *op. cit.* pp. 59–60.
34. Hume, A.O., 1879. Gleanings from the Calcutta market. *Stray Feathers* 7 (6): 479–498 [492–493].
35. BirdLife International, 2001. Pink-headed Duck *Rhodonessa caryophyllacea*. In: Collar, N.J., Andreev, A.V., Chan, S., Crosby, M.J., Subramanya, S. & Tobias, J.A., (eds.). *Threatened Birds of Asia: The BirdLife International Red Data Book*. BirdLife International, Cambridge, UK. Vol. 1. pp. 489–501.
36. Walters, M., 1998. The eggs of the Pink-headed Duck. *Bulletin of the British Ornithologists' Club* 118 (3): 187–191.
37. Sakthivel, R., Dutta, B.B. & Sanyal, A.K., 2011. Catalogue of eggs (Aves) in the National Zoological Collection of the Zoological Survey of India (Part-1). *Records of the Zoological Survey of India, Occasional Paper No.* 325: i–xvi, 1–308.
38. Zarrin, A., 2023. Documenting the bird collection in the State Museum, Lucknow (Uttar Pradesh, India). *Indian BIRDS* 18 (6): 163–180. ['. . . had been mislabelled as a "Spotted billed duck."' (p. 168.)]
39. Reid, G., 1886. *Catalogue of the Birds in the Provincial Museum, N. W. P. & Oudh, Lucknow, on the 1st January 1886*, 1st ed. Museum Committee. Printed by the Calcutta Central Press Company, Limited, Calcutta. pp. i–viii, 1–221.
40. Reid, G., 1890. *Catalogue of the Birds in the Provincial Museum, N.-W. P. and Oudh, Lucknow, on the 1st April 1889*, 2nd ed. North-Western Provinces and Oudh Government Press, Allahabad. pp. i–viii, 1–221.
41. Prestwich, A.A., 1974. The Pink-headed Duck (*Rhodonessa caryophyllacea*) in the wild and in captivity. *The Avicultural Magazine* 80 (2): 47–52.
42. Jerdon, T.C., 1864. *The Birds of India: Being a Natural History of All The Birds Known to Inhabit Continental India; With Descriptions of the Species, Genera, Families, Tribes, and Orders,*

and a Brief Notice of Such Families as Are Not Found in India, Making it a Manual of Ornithology Specially Adapted For India. 1st ed. Published by the author (Printed by George Wyman and Co., Calcutta). pp. i–iv, 2 ll., 441–876, i–xxxii. ['It is excellent eating,' p. 801.]

43. Inglis, C.M., 1940. Records of some rare or uncommon geese, ducks and other water birds in north Bihar. *Journal of the Bengal Natural History Society* XV: 56–60.

44. Hume, J.P., 2018. A high price to pay: New light on the extinction of the Pink-headed Duck *Rhodonessa caryophyllacea. Forktail* 33: 56–63 (2017).

45. Hume, J.P., 2018. *op. cit.*, p. 60.

46. Anonymous, 2023. David Elias Ezra. https://en.wikipedia.org/wiki/David_Elias_Ezra (accessed on 16 July 2024)

47. Hume, J.P., 2018. *op. cit.*, p. 59.

48. Hume, J.P., 2018. *op. cit.*, pp. 59–60.

49. Hume, J.P., 2018. *op. cit.*, p. 62.

50. Ara, J., 1960. In search of the Pinkheaded Duck [*Rhodonessa caryophyllacea* (Latham)]. *Journal of the Bombay Natural History Society* 57 (2): 415–417.

51. Kazmi, R., 2023. The first lady of Indian ornithology. In: Mani, A., (Ed.) *Women in the Wild: Stories of India's Most Brilliant Women Wildlife Biologists.* Juggernaut Books, New Delhi. pp. 13–54.

52. Hume, J.P., 2018. *op. cit.*, p. 33. [Hume erred in the gender.]

53. Nugent, R., 1991. *The Search for the Pink-headed Duck.* 1st ed. Houghton Mifflin Company, Boston. pp. i–xii, 1–223.

54. Eames, J.C., 2004. Northern Myanmar wetland survey. *The Babbler* 12 (December): 9–10.

55. Thorns, R., 2017. Search for the Pink-headed Duck: The interviews. Medium: Lost Species. https://medium.com/@LostSpecies/search-for-the-pink-headed-duck-the-interviews-b576a942e553.

56. Ericson, P.G.P., Qu, Y.H., Blom, M.P.K., Johansson, U. S. & Irestedt, M., 2017. A genomic perspective of the Pink-headed Duck *Rhodonessa caryophyllacea* suggests a long history of low effective population size. *Scientific Reports* 7: 16853.
57. Jerdon, T.C., 1846. *Illustrations of Indian Ornithology*. Reuben Twigg, Christian Knowledge Society's Press, Church Street, Vepery, India. Vol. III of IV vols. pp. 20 ll., pll. XXVI–XL (col.) (April 1846).
58. Ezra, A., 1926. The Pink-headed Duck (*Rhodonessa caryophyllacea*). *The Avicultural Magazine* (4) 4 (12): 325.
59. Seth-Smith, D., 1932. Foxwarren Park. *The Avicultural Magazine* (4) 10 (6): 117–120.
60. Walters, M., 1994. *Birds' Eggs*. 1st ed. Dorling Kindersley, London, New York, Stuttgart. pp. 1–256.
61. Sakthivel et al., 2011. *op. cit.*, p. 20.
62. Ali, S., 1977. *The Book of Indian Birds*. 10th (Revised and enlarged) ed. Bombay Natural History Society, Bombay. pp. i–xlvii, 1–175.
63. Ali, S., 1961. *The Book of Indian Birds*. 6th (Revised and enlarged) ed. Bombay Natural History Society, Bombay. pp. i–xlvi, 1–158, xlvii–li.
64. Nicolai, B., 2013. Vogelmaler und Illustratoren in Mitteleuropa als Mittler zwischen Kunst und Wissenschaft (Bird artists and illustrators in Central Europe as mediators between art and science). *Der Ornithologische Beobachter* 110 (3): 359–372.
65. Dickinson, E., 1891. 'Hope' [254]. In: Johnson, T. H., 1975. *The Complete Poems of Emily Dickinson*. Faber & Faber Ltd., London.
66. Wikipedia, 2024. Evidence of absence. https://en.wikipedia.org/wiki/Evidence_of_absence#:~:text=Evidence%20of%20absence%20and%20absence,a%20writing%20by%20William%20Wright (accessed on 31 July 2024) [The author of this antimetabole has not been traced. It has been used at least since 1881.]

2. The Rediscovery of the Jerdon's or Double-banded Courser

1. Bhushan, B., 1985a. Jerdon's or Double-banded Courser *Cursorius bitorquatus* (Blyth) – Preliminary Survey: Pennar river valley areas. Andhra Pradesh. Technical Report No. 9., Endangered Species Project. Bombay Natural History Society, Bombay.
2. *The Birds of India* (Vol. II, Part II, 1877).
3. *The Birds of India* (Vol. II, Part II, 1877).
4. Elliot, W., 1873. Memoir of Dr. T. C. Jerdon. *History of the Berwickshire Nature Club* 7: 143–151.
5. *The Birds of India* (Vol. II, Part II, 1877).
6. *Journal of the Asiatic Society of Bengal*, 17: 254.
7. *Journal of the Asiatic Society of Bengal*, 18: 260.
8. Blyth, E., 1848. Proceedings of the Asiatic Society – Report of the Curator, Zoological Department. *Journal of the Asiatic Society of Bengal*. xvii (1): 254.
9. In 1896 R.B. Sharpe recognized the genus, *Rhinoptilus*, as described by H.E. Strickland in 1852 in the *Proceedings of the Zoological Society of London* (: 220). Strickland included the *bitorquatus* (the Double-banded Courser) with other congenerics in *Rhinoptilus*.
10. Blanford, W.T., 1898. *The Fauna of British India, including Ceylon and Burma*. Birds, Vol. IV. Taylor & Francis, London.
11. Ripley, S.D., 1952. Vanishing and extinct bird species of India. *Journal of the Bombay Natural History Society* 50: 902–206.
12. A Synopsis of the Birds of India and Pakistan, Together with Those of Nepal, Sikkim, Bhutan and Ceylon; Ripley, Sydney Dillon II:Published by Bombay: Bombay Natural History Society: First edition, 1961.

13. Ali, S. & Ripley, S.D., 1969. *Handbook of the Birds of India and Pakistan*. Vol. 3. Compact Edition. Oxford University Press. pp.11–12.
14. Blanford, W.T., 1898. *The Fauna of British India*. Birds, Vol. IV. Taylor & Francis, London.
15. Ali, S., 1977. President's letter: 'Mystery' birds of India-2: Jerdon's or Double-banded Courser. *Hornbill* 1977 (Oct–Dec): 5–7.
16. Ali, S. & Ripley, S.D., 1969. *Handbook of the Birds of India and Pakistan*. Vol. 3. Compact Edition. Oxford University Press, pp.11–12.
17. Ali, S., 1977. President's letter: 'Mystery' birds of India-2: Jerdon's or Double-banded Courser. *Hornbill* 1977 (Oct–Dec): 5–7.
18. Jerdon, T.C., 1877. *The Birds of India*. Vol. II. Part II. Calcutta. pp. 626–629.
19. Blanford, W.T., 1898. *The Fauna of British India*. Birds. Vol. IV. Taylor & Francis, London.
20. The study compared Jerdon's Courser to eight other glareolid relatives, including seven cursoriines (*Pluvianus aegyptius, Rhinoptilus africanus, R. cinctus, R. chalcopterus, Cursorius cursor, C. temminckii, C. coromandelicus*, and, as an out-group taxon, a single glareoline, *Stiltia isabella*). Based on the cladogram that was developed through the PAUP analysis, it was concluded that the two polytypic genera in the 'cursoriine' assemblage were valid. The two polytypic genera within the 'cursoriinae' were identified as Rhinoptilus and Cursorius, along with a single monotypic genus Pluvianus. Four species were listed by Ripley and Beehler as forming a well-defined clade, that is, Rhinoptilus. These included bitorquatus, cinctus, africanus and chalcopterus. The species cursor, temminckii and coromandelicus, were allied into the genus Cursorius.

21. Ripley, S.D. & Beehler, B.M., 1989. Systematics, biogeography and conservation of Jerdon's Courser *Rhinoptilus bitorquatus*. *Journal of the Yamashina Institute of Ornithology*. 21 (2) 165–174.
22. The Nagari hill tracts and Nallamalai ranges are areas that are 'politically' outside of the Rayalaseema districts of Andhra Pradesh now, but 'historically' they were part of the 'seema' (border) districts of the Krishnadeva 'Raya' kingdom.
23. Jerdon, T.C., 1877. *The Birds of India*. Vol. II. Part II. Calcutta. pp. 626–629.
24. Blanford, W.T., 1898. *The Fauna of British India*. Birds. Vol. IV. Taylor & Francis, London.
25. Bhushan, B., 1994. Ornithology of the Eastern Ghats (in south Andhra Pradesh). Phd Thesis. Bombay Natural History Society, University of Mumbai.

3. A Tale of an Absconding Owl

1. Ali, S. 1948. The Gujarat Satpuras in Indian ornitho-geography. Gujarat Research Society Monographs No. 2: 1-11.
2. Ripley, S.D. 1952. Vanishing and extinct bird species of India. *Journal of the Bombay Natural History Society* 50: 902-906.
3. Ripley, S.D. (1976) Reconsideration of *Athene blewitti* (Hume). *Journal of the Bombay Natural History Society* 73: 1-4.
4. Ali, S. 1978. Mystery birds of India, 3: Blewitt's Owl or Forest Spotted Owlet. Hornbill (Jan-Mar): 4-6.
5. Rasmussen, P.C. and N.J. Collar. 1998. Identification, distribution and status of the Forest Owlet *Athene* (*Heteroglaux*) *blewitti*. *Forktail* 14: 41-49.
6. Knox, A.G. 1993. Richard Meinertzhagen – a case of fraud examined. Ibis 135: 320-325.
7. Rasmussen, P.C. and R. Prys-Jones. 2003. History *vs* mystery: The reliability of museum specimen data. *Bulletin of the British Ornithologists' Club* 123A: 66-94.

8. Rasmussen, P.C. and J.C. Anderton. 2005. *Birds of South Asia: the Ripley Guide*. Lynx Edicions, Barcelona and National Museum of Natural History, Smithsonian Institution, Washington, D.C.
9. Ali, S. and S.D. Ripley. 1987. *Compact Handbook of the Birds of India and Pakistan*. Second Edition. Oxford University Press, Delhi, India, pp. 737.
10. Ibid.
11. Rasmussen, P.C. and N.J. Collar. 1999. Major specimen fraud in the Forest Owlet *Heteroglaux* (*Athene* auct.) *blewitti*. Ibis 141: 11-21.
12. Ripley, S.D. (1976) Reconsideration of *Athene blewitti* (Hume). *Journal of the Bombay Natural History Society* 73: 1-4.
13. Rasmussen, P.C. and N.J. Collar. 1998. Identification, distribution and status of the Forest Owlet *Athene* (*Heteroglaux*) *blewitti*. Forktail 14: 41-49.
14. Rasmussen, P.C. and N.J. Collar. 1999. Major specimen fraud in the Forest Owlet *Heteroglaux* (*Athene* auct.) *blewitti*. Ibis 141: 11-21.
15. Ali, S. and S.D. Ripley. 1969. *Handbook of the Birds of India and Pakistan*. Stone-curlews to owls. First Edition. Oxford University Press, Bombay, India.
16. Knox, A.G. and M.P Walters. 1994. Extinct and Endangered Birds in the Collections of The Natural History Museum, Tring, UK. *British Ornithologists' Club Occasional Publication* No. 1.
17. Rasmussen, P.C. and N.J. Collar. 2013. Phenotypic evidence for the specific and generic validity of *Heteroglaux blewitti*. Forktail 29: 78-87.
18. King, B.F. and P.C. Rasmussen. 1998. The rediscovery of the Forest Owlet *Athene* (*Heteroglaux*) *blewitti*. Forktail 14: 51-53.
19. Rasmussen, P.C. and Ishtiaq, F. 1999 Vocalizations and behaviour of the Forest Owlet Athene (*Heteroglaux blewitti*). Forktail 15: 61-66.

4. The Chilappan Challenge

1. Praveen, J., and P. O. Nameer. 2008. Bird diversity of Siruvani and Muthikulam Hills, Western Ghats, Kerala. *Indian BIRDS* 3(6): 210–217.
2. Jerdon, T. C. 1863. *The Birds of India – Being a Natural History of all the Birds Known to Inhabit Continental India: with Descriptions of the Species, Genera, Families, Tribes, and Orders, and a Brief Notice of such Families as are not Found in India. Making a Manual of Ornithology Specially Adapted for India.* Volume 2. Military Orphan Press, Calcutta, India.
3. Jerdon, T. C. 1839. Catalogue of the birds of the peninsula of India, arranged according to the modern system of classification; with brief notes on their habits and geographical distribution, and description of new, doubtful and imperfectly described species. *Madras Journal of Literature and Science* X (25):234–269.
4. Tobias, J.A., N. Seddon, C.N. Spottiswoode, J.D. Pilgrim, L.D.C. Fishpool, and N.J. Collar (2010). Quantitative criteria for species delimitation. *Ibis* 152(4):724–746.
5. Praveen, J., and P.O. Nameer. 2013. Strophocincla laughingthrushes of south India: A case for allopatric speciation and impact on their conservation. *Journal of the Bombay Natural History Society* 109 (1–2): 46–52.
6. Robin, V.V., C.K. Vishnudas, P. Gupta, F.E. Rheindt, D.M. Hooper, U. Ramakrishnan, and S. Reddy. 2017. Two new genera of songbirds represent endemic radiations from the Shola Sky Islands of the Western Ghats, India. *BMC Evolutionary Biology* 17: 31.
7. Blyth, E. 1851. Notice of a collection of Mammalia, birds, and reptiles, procured at or near the station of Chérra Punji in the Khásia Hills, north of Sylhet. *Journal of the Asiatic Society of Bengal* XX (Part II): 517–524.

8. Rasmussen, P.C., and J.C. Anderton. 2005. *Birds of South Asia. The Ripley Guide. Volumes 1 and 2*. Smithsonian Institution, Washington, D.C., USA and Lynx Edicions, Barcelona, Spain.
9. Monophyly is a scenario where a group of organisms sharing a single common ancestor. Reciprocal monophyly refers to a situation involving two or more groups of organisms where each group has a distinct common ancestor and that ancestor is not shared with any other group.
10. Davison, W. 1883. Notes on some birds collected on the Nilghiris and in parts of Wynaad and southern Mysore. *Stray Feathers* 10: 329–419.
11. BirdLife International. 2024. Species factsheet: *Montecincla jerdoni*. Downloaded from https://datazone.birdlife.org/species/factsheet/banasura-chilappan-montecincla-jerdoni.
12. Reserve forests, managed by the government since the British Raj, differ from vested forests, which were acquired from private parties under the Kerala Private Forests (Vesting & Assignment) Act 1971. Despite court rulings favouring the government, ongoing private claims make it challenging to declare sanctuaries in vested forests.

5. Following in the Footsteps of the Elusive Masked Finfoot

1. Seasonal freshwater wetlands.
2. Oxbow lakes.
3. A lake-like wetland.
4. Jhum cultivation is a traditional farming method that involves slash-and-burn clearing of land in order to grow crops. It is also called shifting cultivation.

5. A group of species (living and extinct) that have a common ancestor.
6. A wetland area of international importance under the Ramsar Convention, also known as 'The Convention on Wetlands', an intergovernmental treaty that provides the framework for the conservation of wetlands. The Convention was adopted in the Iranian city of Ramsar in 1971 and came into force in 1975. Since then, almost 90 per cent of UN member states, from all the world's geographic regions, have become parties to this treaty.
7. https://www.researchgate.net/publication/342709152_Observations_of_the_breeding_of_the_Endangered_Masked_Finfoot_Heliopais_personatus_in_the_Bangladesh_Sundarbans.
8. The Masked Finfoot is absent from the Indian Sundarbans, possibly due to higher water salinity, which may influence their prey (e.g., crab) abundance and associated vegetation structure.

6. Owl of the Emerald Island

1. Abdulali, H., 1964a. The birds of the Andaman and Nicobar Islands. *Journal of the Bombay Natural History Society* 63: 140–190.
2. Abdulali, H., 1967a. More new races of the birds from the Andaman and Nicobar. *Journal of the Bombay Natural History Society* 63: 420–422.
3. Abdulali, H., 1964b. Four new races of the birds from the Andaman and Nicobar islands. *Journal of the Bombay Natural History Society* 61: 410–417.
4. Abdulali, H., 1967b. The birds of Nicobar Islands. *Journal of the Bombay Natural History Society* 64: 139–190.
5. Abdulali, H., 1976. The fauna of Narcondam Island. *Journal of the Bombay Natural History Society* 71: 496–505.

6. Abdulali, H., 1977. New name of Andaman Blackheaded Oriole, *Oriolus xanthornus andamanensis*. *Journal of the Bombay Natural History Society* 73 (2): 395.

7. Blyth, E., 1845. Notices and descriptions of various new and little known species of birds. *Journal of the Asiatic Society of Bengal* 14: 546–602.

8. Blyth, E., 1846a. Notices and descriptions of various new and little known species of birds. *Journal of the Asiatic Society of Bengal* 15: 1–54.

9. Blyth, E., 1846b. Notes on the fauna of the Nicobar Islands. On collections by Mr. Barbe and Capt. Lewis. *Journal of the Asiatic Society of Bengal* 15: 367–379.

10. Blyth, E., 1863. Zoology of Andaman Islands. Appendix (pp. 345–367, with Tytler's notes) to Mouat (1863). *Journal of the Asiatic Society of Bengal* 32: 85–89.

11. Blyth, E., 1866. Abstracts from letters from Capt. Blair. *Ibis* (2): 220–221.

12. Tytler, R.C., 1864. Description of a new species of *Paradoxurus* from the Andaman Islands. *Journal of the Asiatic Society of Bengal*, XXXIII (Part II, Series, No 294, No II): 188.

13. Tytler, R.C., 1867. The avifauna of the Andaman Islands. *Ibis* 3 (2): 314–334.

14. Ball, V., 1870. Notes on birds observed in the neighbourhood of Port Blair, Andaman Islands, during month of August 1864. *Journal of the Asiatic Society of Bengal* 39: 240–243.

15. Ball, V., 1873. List of birds known to occur in the Andaman and Nicobar Islands. *Stray Feathers* I (2–4): 51–90.

16. Ball, V., 1872. Notes on a collection of birds made in Andaman Islands by Asst. Surgeon, G.E. Dobson M.B., during months of April and May. *Journal of the Asiatic Society of Bengal* 41: 273–290.

17. Hume, A.O., 1873a. Additional remarks on the avifauna of the Andamans. *Stray Feathers* I (2–4): 304–310.

18. Hume, A.O., 1873b. Notes. Avifauna of the islands of the Bay of Bengal. *Stray Feathers* 5: 421–423.
19. Hume, A.O., 1874a. Additional notes on the avifauna of the Andaman Islands. *Stray Feathers* 2 (6): 490–501.
20. Hume, A.O., 1874b. Contributions to the ornithology of India. The Islands of the Bay of Bengal. *Stray Feathers* 2 (1–3): 29–324.
21. Hume, A.O., 1876. Additional notes on the avifauna of the Andaman Islands. *Stray Feathers* 4 (4–6): 279–294.
22. Butler A.L. 1899a. The birds of the Andaman and Nicobar Islands. Part I. *Journal of the Bombay Natural History Society* 12(2):386–403.
23. Butler, A.L., 1899b. The birds of the Andaman and Nicobar Islands. Part II. *Journal of the Bombay Natural History Society* 12 (3): 555–571.
24. Butler, A.L., 1899c. The birds of the Andaman and Nicobar Islands. Part III. *Journal of the Bombay Natural History Society* 12 (4): 684–696.
25. Butler, A.L., 1900. The birds of the Andaman and Nicobar Islands. Part IV. *Journal of the Bombay Natural History Society* 13 (1): 144–154.
26. Richmond, C.W., 1902. Birds collected by Dr. W. L. Abbot and Mr. C. B. Kloss in the Andaman and Nicobar Islands. *Proceedings of the United States National Museum* 25: 287–314.
27. Rasmussen, P.C., 1998. A new Scops Owl from Great Nicobar Island. *Bulletin of the British Ornithologists' Club* 118 (3): 141–153.
28. Ali, S & Ripley, S.D., 1983. *Handbook of the birds of India and Pakistan*, Compact edition. Oxford.
29. Ripley, S.D. & Beehler, B.M., 1989. Ornithographic affinities of the Andaman and Nicobar Islands. *Journal of Biogeography* 16 (4): 323–332.

30. Sankaran, R. 1998. An annotated list of the endemic avifauna of the Nicobar Islands. *Forktail* 13 (February): 17–22. University Press, New Delhi. 17–22.
31. Incidentally, Richmond just wrote about the species in a scientific paper and never set foot on the Nicobars. It was Abbott and Kloss who collected specimens, which is why some species and subspecies of the islands are named after them, including the Great Nicobar subspecies of Nicobar Megapode, Nicobar Pitta, Great Nicobar Serpent Eagle, Ornate Sunbird and the Andaman subspecies of Red-breasted Parakeet.

7. In Search of the Last Megapodes

1. The Nicobar Megapode occurs as two distinct subspecies – *Megapodius nicobariensis* in the Nancowry group of islands and *Megapodius nicobariensis abbotti* in the Great Nicobar group. Historically it occurred on most Nicobar Islands except Car Nicobar and Bati Malv. There were a few records from the Andaman group of Islands, however, the species is now believed to be locally extinct there. (Sivakumar, K., 2000) A study on the breeding biology of the Nicobar megapode Megapodius nicobariensis. PhD Thesis, Bharathiyar University, Coimbatore, India).
2. Fernández-Palacios, et al. 2021. Scientists' warning – The outstanding biodiversity of islands is in peril. *Global Ecology and Conservation*, 31, Article e01847.
3. There are several theories on the origin of megapodes. Millions of years ago, the family probably had a common ancestor that lived on a supercontinent. During the Ice Age (Pleistocene), when the sea levels were lowered, Australia had a land connection with far away islands such as New Guinea and Japan. This land connection existed as recently as 20,000 years

ago. When the ice began to melt leading to a rise in sea levels, several low-lying areas were submerged, leaving behind a string of isolated islands. Wherever large predators were absent, the megapodes flourished. Gradually they evolved in sync with their surroundings, diverging into distinct species. Jones, D.N., Dekker, R. W. R. J., & Roselaar, C.S. (1995). The Megapodes: Megapodiidae. Oxford [England]: Oxford University Press.

4. Turkeys, megapodes and chickens all belong to the order Galliform, or heavy-bodied ground-feeding birds. Jones, D.N., Dekker, R. W. R. J., & Roselaar, C.S. 1995. The Megapodes: Megapodiidae. Oxford [England]: Oxford University Press.)

5. Hume, A.O., & Marshall, C.H.T. 1878. *The Game Birds of India, Burmah, and Ceylon* (Vol. 1, p. 119). London: A.K.H. Boyd.

6. At the former wildlife camp '41', seven of the eight people there died during the Tsumani. Sankaran R., Andrews H. & Vaughan A. 2005. The Ground Beneath the Waves: Post-tsunami Impact Assessment of Wildlife and their Habitats in India. Vol-2. Wildlife Trust of India, New Delhi, India.

7. Sankaran, R.; Sivakumar, K. 1999. Preliminary Results of an Ongoing Study of the Nicobar Megapode Megapodius Nicobariensis Blyth. Zoologische Verhandelingen.

8. Sankaran R., Andrews H. & Vaughan A. 2005. The Ground Beneath the Waves: Post-tsunami Impact Assessment of Wildlife and their Habitats in India. Vol-2. Wildlife Trust of India, New Delhi, India. 105 p.

9. Sivakumar, K. (2010). *Impact of the 2004 Tsunami on the Vulnerable Nicobar Megapode Megapodius nicobariensis*. Oryx 44(1): 71-78.

10. Estimate from a 2006 survey shows that 395–790 breeding pairs of the Nicobar megapode now survive on the coasts of the various islands compared to 2,318–4,056 pairs that

Sivakumar estimated in 1994. (Sivakumar, K., 2010). *Impact of the 2004 tsunami on the Vulnerable Nicobar Megapode Megapodius nicobariensis. Oryx* 44(1): 71-78.
11. Sekhsaria, P. 2019. Islands in Flux – the Andaman and Nicobar Story (2nd ed.). HarperCollins Publishers India.
12. Sekhsaria, P. 2015. Disaster as a Catalyst for Military Expansionism: The Case of the Nicobar Islands. *Economic and Political Weekly*. Vol. 50, No. 1 (3 January 2015), pp. 37-43.
13. In October 2022, the Andaman and Nicobar administration issued official notifications declaring three wildlife sanctuaries – Leatherback Turtle Sanctuary at Little Nicobar Island, Coral Sanctuary at Meroe Island and Megapode Sanctuary at Menchal Island.

8. Nong-in – The Bird that Tracks the Rain

1. Morung Express (n.d.). Jessami revisited. https://morungexpress.com/jessami-revisited (accessed on 21 December 2023).
2. Britain's Greatest Battles (n.d.). Online Exhibitions | National Army Museum, London. https://web.archive.org/web/20131225171213/http://www.nam.ac.uk/exhibitions/online-exhibitions/britains-greatest-battles (accessed on 28 August 2024).
3. Yaiphaba, S. Nong-in, the state bird of Manipur. e-pao.net (accessed on 21 December 2023).
4. Hume, A.O., 1880. *Stray Feathers*, 9, 461–467. https://www.biodiversitylibrary.org/page/29834800 (accessed on 21 December 2023).
5. The Meiteis are the largest ethic group in the state of Manipur, and the Meitei language is one of India's official languages.
6. Hume, A. O. 1880. *Stray Feathers*, 9, 461–467. https://www.biodiversitylibrary.org/page/29834800 (accessed on 21 December 2023).

7. *Ibid.*
8. *Ibid.*
9. *Ibid.*
10. Spencer, A.J., Sharma, P., Huidrom, H., Laishram, D. & Sharma, K.J., 2022. Grey-eyed Bulbul Iole propinqua from Manipur, India: An addition to the avifauna of South Asia, with notes on its vocalisations. *Indian BIRDS* 18–4: 113–116. https://indianbirds.in/pdfs/IB_18_4_SpencerETAL_GreyeyedBulbul.pdf (accessed on 28 August 2024).
11. Sharma, P., Spencer, A.J., Laishram, D., Akoijam, P., Huidrom, H., Sharma, K.J. & Singh. 2022. Rufous-winged Buzzard Butastur liventer from Manipur, India: An addition to the avifauna of South Asia. *Indian BIRDS* 18–3: 86-88. https://indianbirds.in/pdfs/IB_18_3_SharmaETAL_RufouswingedBuzzard.pdf (accessed on 28 August 2024).
12. Choudhury, A., 2005. New sites for Mrs Hume's Pheasant *Syrmaticus humiae* in north-east India based on hunters' specimens and local reports. *Forktail* 21:183. https://static1.squarespace.com/static/5c1a9e03f407b482a158da87/t/5c1ffae803ce64637dbe819b/1545599720658/Choudhury-Humes.pdf (accessed on 28 August 2024).
13. Ukhrul Times, 2024. Mrs Hume's Pheasant Community Reserve opens in Jessami. https://ukhrultimes.com/mrs-humes-pheasant-community-reserve-opens-in-jessami/ (accessed on 28 August 2024).
14. Mukherjee, S. 2024. Mrs. Hume's Pheasant – WildArt.Works. WildArt.Works. https://wildart.works/behindthelens/mrs-humes-pheasant#:~:text=The%20state%20bird%20of%20Manipur%2C%20Mrs.,long%2Dtailed%20terrestrial%20forest%20pheasant (accessed on August 28 2024).

15. Singh, R.K. 2009. Where are you 'Nongin'? – The State Bird of Manipur Part 1. https://e-pao.net/. Retrieved on 28 August 2024, from https://e-pao.net/epSubPageExtractor.asp?src=education.Science_and_Technology.Where_are_you_Nongin_1.
16. *Ukhrul Times*, 2024. Mrs Hume's Pheasant Community Reserve opens in Jessami. https://ukhrultimes.com/mrs-humes-pheasant-community-reserve-opens-in-jessami/. (accessed on 28 August 2024).

9. In Search of Vicky on Phawngpui

1. Sharma, P., Laishram, D. & Huidrom, H., 2021. Chestnut-crowned Bush Warbler *Cettia major*: First records for Manipur, India, and some notes on its vocalisations. *Indian BIRDS* 17 (6): 179–181.
2. There are eleven genera of Bulbuls (*Pycnonotidae*) currently recognized in South Asia, including the genus *Iole*, which consists of rather uniform and subdued species that, in contrast to many others in their family, tend to be inconspicuous and often go unnoticed. In north-eastern India, the genus was represented by a single species, Cachar Bulbul (*Iole cacharensis*), which was until recently recognized as a subspecies of the Olive Bulbul (*I. viridescens*).
3. Spencer, A.J., Sharma, P., Huidrom, H., Laishram, D. & Sharma, J., 2022. Grey-eyed Bulbul *Iole propinqua* from Manipur, India: An addition to the avifauna of South Asia, with notes on its vocalisations. *Indian BIRDS* 18 (4): 113–116.
4. Sharma, P., Spencer, A.J., Laishram, D., Akoijam, P., Huidrom, H., Sharma, K.J. & Singh, S.O., 2022. Rufous-winged Buzzard *Butastur liventer* from Manipur, India: An addition to the avifauna of South Asia. *Indian BIRDS* 18 (3): 86–88.

5. Spencer, A. & Sharma, P., 2022. eBird Trip Report: Kwatha area, Manipur. 28 Dec 2021–1 Jan 2022. https://ebird.org/tripreport/76348 (accessed on 15 October 2024).
6. *ibid*.
7. *ibid*.
8. Koelz, W., 1954. Ornithological studies. I. New birds from Iran, Afghanistan, and India. *Contributions from the Institute for Regional Exploration* 1: 1–32.
9. Ghose, D., 1999a. Phawngpui, home of Blyth's Tragopan in Mizoram, India. *Twilight* 1 (5): 14–15.
10. Ghose, D., 1999b. Birds recorded at Blue Mountain (Phawngpui) National Park, Mizoram, between February–May, 1997. *Twilight* 1 (5): 16–18.
11. Singh, P. 2016. Sound-recording. https://xeno-canto.org/748584 (accessed on 15 October 2024).
12. Bird Count India, 2021. Fantastic birds and where to find them. https://birdcount.in/rarest-birds-of-india/ (accessed on 15 October 2024).
13. Spencer, A. & Sharma, P., 2022. eBird trip report: Mizoram 2022. https://ebird.org/tripreport/32829 (accessed on 15 October 2024).
14. *Hindustan Times*, 2022. Mount Victoria Babax bird species spotted in India after 25 years. https://www.hindustantimes.com/india-news/mount-victoria-babax-bird-species-spotted-in-india-after-25-years-101642187110286.html (accessed on 15 October 2024).
15. Bird Count India, 2022. Finding Vicky on Phawngpui! https://birdcount.in/mount victoria-babax/ (accessed on 15 October 2024).

10. Lusting for a *Locustella*

1. Biddulph, J. 1881. The birds of Gilgit. *Stray Feathers* 9: 301–366.
2. Buchanan, K. 1903. Nesting notes from Kashmir. *Journal of the Bombay Natural Historical Society* 15: 131–133.
   Davidson, J. 1898. A short trip to Kashmir. *Ibis* 4 (7): 1–42.
3. Kennerley, P. & Pearson, D., 2010. *Reed and bush warblers*. A. & C. Black.
4. Anonymous. 1979. University of Southampton Himalayan Expedition 1977 report. University of Southampton, UK.
5. BirdLife International, 2022. Species factsheet: Locustella major. http://www.birdlife.org (accessed on 5 October 2022).
6. Re:Wild (2022) https://www.rewild.org/news/four-re-wild-initiatives-to-follow-in-2022-four-rewilding-wins-to-celebrate.

11. Twitching Tales

1. Jatinga is a village in southern Assam that has earned some notoriety for birds literally falling out of the sky in certain months of the year, only to be nabbed by waiting hunters. The rational explanation for the phenomenon is that the inexperienced juvenile birds, and those that have locally migrated from another area, may be disoriented by the high-velocity monsoon winds along with fog typical of the monsoon season. In their dazed state, the birds automatically fly towards sources of light, hence the impression that birds are falling out of the sky. Of course, what awaits them is not rescue – incoming birds often get a hard thwack from a bamboo pole and are collected (presumably) for the pot.

List of Non-human Animals Mentioned in the Book
(Scientific names based on IOC taxonomy)

Introduction

1. Mishmi Wren-Babbler *Spelaeornis badeigularis*
2. Nicobar Scops Owl *Otus alius*
3. Bugun Liocichla *Liocichla bugunorum*
4. Himalayan Forest Thrush *Zoothera salimalii*
5. Ashambu Sholakili *Sholicola ashambuensis*
6. Forest Owlet *Athene blewitti*
7. Pink-headed Duck *Rhodonessa caryophyllacea*
8. Mount Victoria Babbax *Pterorhinus woodi*
9. Masked Finfoot *Heliopais personatus*
10. Banasura Laughingthrush *Montecincla jerdoni*
11. Jerdon's Courser *Rhinoptilus bitorquatus*
12. Mrs Hume's Pheasant *Syrmaticus humiae*
13. Yellow-throated Laughingthrush *Pterorhinus galbanus*
14. Brown-capped Laughingthrush *Trochalopteron austeni*
15. Moustached Laughingthrush *Ianthocincla cineracea*
16. Gould's Shortwing *Heteroxenicus stellatus*
17. Rusty-bellied Shortwing *Brachypteryx hyperythra*
18. Hodgson's Frogmouth *Batrachostomus hodgsoni*
19. Sikkim Wedge-billed Babbler *Stachyris humei*
20. Cachar Wedge-billed Babbler *Stachyris roberti*

1. A Charismatic Duck Painted in Carnation and Chocolate

1. Pink-headed Duck *Rhodonessa caryophyllacea*
2. Red-crested Pochard *Netta rufina*
3. Himalayan Quail *Ophrysia superciliosa*
4. Jerdon's Courser *Rhinoptilus bitorquatus*
5. Forest Owlet *Athene blewitti*
6. Indian Rhinoceros *Rhinoceros unicornis*

2. The Rediscovery of the Jerdon's or Double-banded Courser

1. Jerdon's Courser *Rhinoptilus bitorquatus*
2. Great Indian Bustard *Ardeotis nigriceps*
3. Lesser Florican *Sypheotides indicus*
4. Red-wattled Lapwing *Vanellus indicus*
5. Golden Gecko *Calodactylodes aureus*

3. A Tale of an Absconding Owl

1. Forest Owlet *Athene blewitti*
2. Pink-headed Duck *Rhodonessa caryophyllacea*
3. Jerdon's Courser *Rhinoptilus bitorquatus*
4. Mountain Quail/Himalayan Quail *Ophrysia superciliosa*
5. Spotted Owlet *Athene brama*
6. Mottled Wood-Owl *Strix ocellata*

4. The Chilappan Challenge

1. Banasura Chilappan/Banasura Laughingthrush *Montecincla jerdoni*
2. Nilgiri Chilappan/Rufous-breasted Laughingthrush/Nilgiri Laughingthrush *Montecincla cachinnans*
3. Palani Chilappan *Montecincla fairbanki*
4. Ashambu Chilappan *Montecincla meridionalis*

5. Nilgiri Sholakili *Sholicola major*
6. Black-and-orange Flycatcher *Ficedula nigrorufa*
7. Rufous Treepie *Dendrocitta vagabunda*
8. Jungle Babbler *Argya striata*

5. Following in the Footsteps of the Elusive Masked Finfoot

1. Masked Finfoot *Heliopais personatus*
2. Spoon-billed Sandpiper *Calidris pygmaea*
3. Great Hornbill *Buceros bicornis*
4. Cotton Pygmy Goose *Nettapus coromandelianus*
5. Pink-headed Duck *Rhodonessa caryophyllacea*
6. Indian Peafowl *Pavo cristatus*
7. Mangrove Pitta *Pitta megarhyncha*
8. African Finfoot *Podica senegalensis*
9. Sungrebe/American Finfoot *Heliornis fulica*
10. Shikra *Tachyspiza badia*
11. Changeable Hawk Eagle *Nisaetus cirrhatus*
12. Sumatran Rhinoceros *Dicerorhinus sumatrensis*
13. Striped Hyena *Hyaena hyaena*
14. Blackbuck *Antilope cervicapra*
15. Gray Wolf *Canis lupus*
16. Nilgai *Boselaphus tragocamelus*
17. Western Hoolock Gibbon *Hoolock hoolock*
18. Elongated Tortoise *Indotestudo elongata*
19. Asiatic Wild Dog/Dholes *Cuon alpinus*
20. Clouded Leopard *Neofelis nebulosa*
21. Asian Elephant *Elephas maximus*
22. Bengal Tiger *Panthera tigris*
23. Spotted Deer *Axis axis*

6. Owl of the Emerald Island

1. Nicobar Scops Owl *Otus alius*
2. Pied Harrier *Circus melanoleucos*

List of Non-human Animals

3. House Crow *Corvus splendens*
4. House Sparrow *Passer domesticus*
5. Common Myna *Acridotheres tristis*
6. Nicobar Imperial Pigeon *Ducula nicobarica*
7. Black Baza *Aviceda leuphotes* (subspecies *andamanica*)
8. Slaty-breasted Rail *Lewinia striata* (subspecies *nicobariensis*)
9. White-breasted Waterhen *Amaurornis phoenicurus* (subspecies *midnicobaricus* from Central Nicobar, and subspecies *leucocephalus* from Car Nicobar)
10. Andaman Cuckoo-Dove *Macropygia rufipennis* (subspecies *tiwarii*)
11. Brown Hawk Owl *Ninox scutulata* (subspecies *rexpimenti*)
12. Asian Fairy Bluebird *Irena puella* (subspecies *andamanica*)
13. Glossy Starling *Aplonis panayensis* (subspecies *albiris*)
14. Black-hooded Oriole *Oriolus xanthornus* (subspecies *reubeni*)
15. Oriental Scops Owl *Otus sunia nicobaricus*
16. Sooty Babbler *Stachyris herberti*
17. Simeulue Scops Owl *Otus umbra*
18. Enggano Scops Owl *Otus enganensis*
19. Pied Imperial Pigeon *Ducula bicolor*
20. Chinese Pond Heron *Ardeola bacchus*
21. Chinese Sparrowhawk *Tachyspiza soloensis*
22. Nicobar Hooded Pitta *Pitta abbotti*
23. Nicobar Parakeet *Psittacula caniceps*
24. Nicobar Jungle Flycatcher *Cyornis nicobaricus*
25. Great Nicobar Serpent Eagle *Spilornis klossi*
26. Nicobar Imperial Pigeon *Ducula nicobarica*
27. Crimson Sunbird *Aethopyga siparaja* (endemic subspecies: *nicobarica*)
28. Olive-backed Sunbird *Cinnyris jugularis* (endemic subspecies: *klossi*)
29. Indian White-eye *Zosterops palpebrosus* (endemic subspecies: *nicobaricus*)
30. Oriental Dwarf Kingfisher *Ceyx erithaca* (subspecies: *macrocarus*)

31. Arctic Warbler *Phylloscopus borealis*
32. Daurian/Purple-backed Starling *Agropsar sturninus*
33. Great Nicobar Crake *Rallina* species
34. Nicobar Megapode *Megapodius nicobariensis*
35. Nicobar Treeshrew *Tupaia nicobarica*
36. Crab-eating Macaque *Macaca fascicularis*
37. Narrow-headed Frog *Microhyla chakrapanii*
38. Cricket Frog *Minervarya andamanensis*
39. *Hemidactylus* species
40. Indian Bullfrog *Hoplobatrachus tigerinus*
41. Olive Ridley Turtle *Lepidochelys olivacea*
42. Green Turtle *Chelonia mydas*

7. In Search of the Last Megapodes

1. Nicobar Megapode *Megapodius nicobariensis*
2. Sri Lankan Frogmouth *Batrachostomus moniliger*
3. Melanesian Megapode *Megapodius eremita*
4. Great Nicobar Parakeet *Psittacula caniceps*
5. Western Tragopan *Tragopan melanocephalus*
6. Nicobar Treeshrew *Tupaia nicobarica*
7. Leatherback Turtle *Dermochelys coriacea*
8. Asian Water Monitor Lizard *Varanus salvator*
9. Robber Crabs *Birgus latro*
10. Pnar Cave Mahaseer *Neolissochilus pnar*

8. Nong-in – The Bird that Tracks the Rain

1. Mrs Hume's Pheasant *Syrmaticus humiae*
2. Elliot's Pheasant *Syrmaticus ellioti*
3. Green Peafowl *Pavo muticus*
4. Black-breasted Parrotbill *Paradoxornis flavirostris*
5. Grey-eyed Bulbul *Iole propinqua*
6. Rufous-winged Buzzard *Butastur liventer*
7. Eurasian Jay *Garrulus glandarius*

8. Hodgson's Frogmouth *Batrachostomus hodgsoni*
9. Mountain Bamboo Partridge *Bambusicola fytchii*
10. Yellow-throated Laughingthrush *Pterorhinus galbanus*
11. Striped Laughingthrush *Trochalopteron virgatum*
12. White-browed Laughingthrush *Pterorhinus sannio*
13. Khalij Pheasant *Lophura leucomelanos*
14. Red Junglefowl *Gallus gallus*
15. Cheer Pheasant *Catreus wallichii*
16. Moustached Laughingthrush *Ianthocincla cineracea*
17. Brown-capped Laughingthrush *Trochalopteron austeni*
18. Spot-breasted Parrotbill *Paradoxornis guttaticollis*
19. Crested Finchbill *Spizixos canifrons*
20. Flavescent Bulbul *Pycnonotus flavescens*

9. In Search of Vicky on Phawngpui

1. Mount Victoria Babax *Pterorhinus woodi*
2. Striated Grassbird *Megalurus palustris*
3. Manipur Bush-Quail *Perdicula manipurensis*
4. Chestnut-crowned Bush Warbler *Cettia major*
5. Grey-eyed Bulbul *Iole propinqua*
6. Rufous-winged Buzzard *Butastur liventer*
7. Eurasian Jay *Garrulus glandarius*
8. Collared Falconet *Microhierax caerulescens*
9. Chestnut-bellied Nuthatch *Sitta cinnamoventris*
10. Burmese Nuthatch *Sitta neglecta*
11. Chinese Babax *Pterorhinus lanceolatus*
12. Striped Laughingthrush *Trochalopteron virgatum*
13. Buff-throated Warbler *Phylloscopus subaffinis*
14. Crested Finchbill *Spizixos canifrons*
15. Eyebrowed Thrush *Turdus obscurus*
16. Gray-sided Thrush *Turdus feae*
17. Stripe-breasted Woodpecker *Dendrocopos atratus*
18. Slender-billed Scimitar-Babbler *Pomatorhinus superciliaris*
19. Spot-breasted Laughingthrush *Garrulax merulinus*

20. Hume's Treecreeper *Certhia manipurensis*
21. Chin Hills Wren-Babbler *Spelaeornis oatesi*
22. Black-throated Parrotbill *Suthora nipalensis*
23. Mountain Bamboo Partridge *Bambusicola fytchii*
24. Himalayan Griffon *Gyps himalayensis*
25. Aberrant Bush-Warbler *Horornis flavolivaceus*
26. Fire-tailed Sunbird *Aethopyga ignicauda*
27. Siberian Crane *Leucogeranus leucogeranus*
28. Wood Duck *Aix sponsa*
29. Olive Bulbul *Iole viridescens*

10. Lusting for a *Locustella*

1. Long-billed Bush Warbler *Locustella major*
2. Flores Scops Owl *Otus alfridi*
3. Red-legged Crake *Rallina fasciata*
4. Boano Monarch *Symposiachrus boanensis*
5. Black-browed Babbler *Malacocincla perspicillata*
6. Himalayan Quail *Ophrysia superciliosa*
7. Himalayan Rubythroat *Calliope pectoralis*
8. Mountain Chiffchaff *Phylloscopus sindianus*
9. Blyth's Rosefinch *Carpodacus grandis*
10. Brooks's Leaf Warbler *Phylloscopus subviridis*
11. Himalayan Ibex *Capra Sibirica Hemalayanus*
12. Markhor *Capra falconeri*
13. Flores Giant Rat *Papagomys armandvillei*

11. Twitching Tales

1. White Tern *Gygis alba*
2. White-spotted Fantail *Rhipidura albogularis*
3. Western Tragopan *Tragopan melanocephalus*
4. Yellow-rumped Flycatcher *Ficedula zanthopygia*
5. Plain Leaf Warbler *Phylloscopus neglectus*
6. Jerdon's Courser *Rhinoptilus bitorquatus*

7. Chin Hills Wren-Babbler *Spelaeornis oatesi*
8. Mrs Hume's Pheasant *Syrmaticus humiae*
9. Black-crowned Scimitar Babbler *Pomatorhinus ferruginosus*
10. Rusty-fronted Barwing *Actinodura egertoni*
11. Blue-winged Minla *Actinodura cyanouroptera*
12. Flavescent Bulbul *Pycnonotus flavescens*
13. Grey Sibia *Heterophasia gracilis*
14. Assam Laughingthrush *Trochalopteron chrysopterum*
15. Narcondam Hornbill *Rhyticeros narcondami*
16. Hodgson's Frogmouth *Batrachostomus hodgsoni*
17. Bugun Liocichla *Liocichla bugunorum*
18. Silver-eared Mesia *Leiothrix argentauris*
19. Pale Blue Flycatcher *Cyornis unicolor*
20. Blue-winged Laughingthrush *Trochalopteron squamatum*
21. Sikkim Wedge-billed Babbler *Stachyris humei*
22. Eyebrowed Wren-Babbler *Napothera epilepidota*
23. Golden Babbler *Cyanoderma chrysaeum*
24. Sapphire Flycatcher *Ficedula sapphira*
25. Yellow-bellied Warbler *Abroscopus superciliaris*
26. Pale-headed Woodpecker *Gecinulus grantia*
27. White-hooded Babbler *Gampsorhynchus rufulus*
28. Mandarin Duck *Aix galericulata*
29. Temminck's Tragopan *Tragopan temminckii*
30. Crimson-browed Finch *Carpodacus subhimachalus*
31. Emerald Dove *Chalcophaps indica*

12. Finding the Next Rare Bird for India

1. Ashambu Sholakili *Sholicola ashambuensis*
2. Himalayan Forest Thrush *Zoothera salimalii*
3. Bugun Liocichla *Liocichla bugunorum*
4. Spotted Flycatcher *Muscicapa striata*
5. Eastern Yellow Wagtail *Motacilla tschutschensis*
6. Two-barred Warbler *Phylloscopus plumbeitarsus*
7. Western Yellow Wagtail *Motacilla flava*

8. Eastern Crowned Warbler *Phylloscopus coronatus*
9. Band-bellied Crake *Zapornia paykullii*
10. Radde's Warbler *Phylloscopus schwarzi*
11. Mount Victoria Babax *Pterorhinus woodi*
12. Victoria Nuthatch/White-browed Nuthatch *Sitta victoriae*
13. Yunnan Fulvetta *Alcippe fratercula*
14. Burmese Bushlark/Burmese Myna *Acridotheres burmannicus*

Acknowledgements

Putting together *The Search for India's Rarest Birds* has been such a rewarding journey. Information and ideas that shaped the book were seeded in conversations that Shashank began years ago with James Eaton and Frank Rheindt, who've long been on their own quests to find rare birds across South and Southeast Asia. Many of the authors in this book shared insights and suggestions over and above their own essays, which were immensely helpful.

A story straight from the source has a certain edge. So, having these essays on avian re-discoveries written by the discoverers themselves makes this book special; there is an immediacy, a richness of detail that draws the reader in. Many thanks are due to all our authors because they snatched time from research projects, doctoral studies and tight work schedules to write these essays, and then patiently worked with us through the edit process. No detail was too small, and for this we are grateful.

Thanks are also due to Nishtha, Wesley and the team at Juggernaut who put together the book you hold in your hands! We were delighted when we learnt that we could include a colour insert within this book. After all, when one reads about such magnificent birds, it is a shame not to see them. Dhritiman Mukerjee, James Eaton and

Andrew Spencer generously shared their images to make this happen.

Shashank would like to express his heartfelt gratitude to his better half, Vishnupriya. She has been a constant support, patiently listening to his ideas (a lot of bird talk) and reading through multiple drafts of this book. A wildlife biologist herself, Vishnupriya has, whether willingly or not, been part of numerous bird rediscoveries, often accompanying her over-enthusiastic husband on field expeditions.

And lastly, Anita would like to thank her spouse, Sukumar, for his support – not on the field but on the (edit) screen! His suggestions have improved the book which we hope you have enjoyed reading.

A Note on the Editors

Shashank Dalvi is a researcher and conservationist with more than 25 years of field experience and immense knowledge of India's natural history. His work has led to the discovery of a new bird species in India, only the fourth such discovery since independence. Shashank has also been instrumental in the conservation of Amur Falcons in Nagaland, for which he won the Carl Zeiss Conservation Award in 2014. His recent work focuses on understanding the impact of biogeographic barriers on the population genetics of birds.

Anita Mani lives, works and birds in Delhi, from where she runs Indian Pitta, her book imprint with Juggernaut Books. In addition to editing books about birds and natural history, Anita writes on technology and communications, a throwback to the time she ran the operations of a communications software company. Her work journey is similar to that of a migrating bird – she has oscillated from writing (first for the *Hindu Business Line*, and later *Business Standard*) to a corporate career and back to writing. For several years, she ran a news and current affairs publication for children called *Child Friendly News*. For now, she is content to watch, read and write about birds.

About Indian Pitta

Indian Pitta is India's first dedicated book imprint for bird and nature lovers, conservationists and policymakers. Our books about wildlife and natural history go beyond field/identification guides, to explore the bigger mosaic of habitats, ecosystems and human interactions that touch the lives of birds and animals. Successful conservation programmes, troubling environmental challenges, personal exploration of a landscape, deep dives into the ecology of a species, the quest for a rare species and the sheer joy of birding – these are some of the ideas that you can expect to explore within the pages of our books.

Also Available

ISBN: 978-93-5345-181-3   ISBN: 978-93-5345-164-6   ISBN: 978-93-5345-415-9
Price: 499/-   Price: 1299/-   Price: 599/-

Copyright Acknowledgements

A Charismatic Duck Painted in Carnation and Chocolate written by **Aasheesh Pittie**

The Rediscovery of the Jerdon's or Double-banded Courser written by **Bharat Bhushan**

A Tale of an Absconding Owl written by **Pamela C. Rasmussen**

The Chilappan Challenge written by **Praveen J.**

Following in the Footsteps of the Elusive Masked Finfoot written by **Sayam U. Chowdhury**

Owl of the Emerald Island written by **Shashank Dalvi**

In Search of the Last Megapodes written by **Radhika Raj**

Nong-in – The Bird that Tracks the Rain written by **Anita Mani and Shashank Dalvi**

In Search of Vicky on Phawngpui written by **Puja Sharma and Andrew Spencer**

Lusting for a *Locustella* written by **James Eaton**

Twitching Tales written by **Atul Jain**

Finding the Next Rare Bird for India written by **Frank E. Rheindt**